EPA 260-R-13-001
February 2013

Toxic Chemical Release Inventory Reporting Forms and Instructions

Revised 2012 Version

Section 313
of the Emergency Planning and Community Right-to-Know Act
(Title III of the Superfund Amendments and Reauthorization Act of 1986)

This Page Intentionally Left Blank

Table of Contents

Examples

Figures

Tables

Appendices

List of Acronyms

ARA	Annual Reportable Amount
BIA	Bureau of Indian Affairs
CAS	Chemical Abstract Services
CBI	Confidential Business Information
CDX	Central Data Exchange
CERCLA	Comprehensive Environmental Response, Compensation, and Liability Act
CFR	Code of Federal Regulations
D&B	Dun & Bradstreet
DPC	Data Processing Center
DQA	Data Quality Alert
EBDCs	Ethylenebisdithiocarbamic Acid, Salts and Esters
eFDP	Electronic Facility Data Profile
EPA	Environmental Protection Agency
EPCRA	Emergency Planning and Community Right to Know Act
ESA	Electronic Signature Agreement
FDP	Facility Data Profile
FIPS	Federal Information Processing Standard
FR	Federal Register
GOCO	Government-Owned, Contractor-Operated
IARC	International Agency for Research and Cancer
ICR	Information Collection Request
MSDS	Material Safety Data Sheets
NA	Not Applicable
NDC	Non-Technical Data Changes
NAICS	North American Industry Classification System
NON	Notice of Non-Compliance
NOSE	Notice of Significant Error
NOTE	Notice of Technical Errors
NPDES	National Pollutant Discharge Elimination System
NTP	National Toxicology Program
OMB	Office of Management and Budget
OSHA	Occupational Safety and Health Act
P2	Pollution Prevention
PACs	Polycyclic Aromatic Compounds
PBBs	Polybrominated Biphenyls
PBT	Persistent Bioaccumulative Toxic
PCBs	Polychlorinated Biphenyls
POTW	Publicly Owned Treatment Works
PPA	Pollution Prevention Act
RCRA	Resource Conservation and Recovery Act
RY	Reporting Year
SBREFA	Small Business Regulatory Enforcement Fairness Act
SIC	Standard Industrial Classification
TDX	TRI Data Exchange
TRI	Toxics Release Inventory
TRIFID	Toxics Release Inventory Facility Identification Number
TRIPS	Toxics Release Inventory Processing System
UIC	Underground Injection Control
USC	United States Code
VOCs	Volatile Organic Compounds

Important Information for Reporting Year 2012

New Information for Reporting Year 2012

Please note that this 2012 version of the Toxics Release Inventory (TRI) Reporting Forms and Instructions document supersedes previous versions of the document.

- **Administrative Stay Lifted for TRI Hydrogen Sulfide (H$_2$S) Reporting**

  A Federal Register notice was issued on Monday, October 17, 2011 (76 FR 64022) announcing that EPA is lifting the Administrative Stay of the EPCRA section 313 toxic chemical release reporting requirements of hydrogen sulfide - H$_2$S - (CAS No. 7783-06-4). Reporting for H$_2$S is required for RY 2012 reporting (i.e., for Form R and Form A Certification Statements that are due to the Agency at midnight on July 1, 2013). For more information about the lifting of the administrative stay for H$_2$S, see: http://www.epa.gov/tri/lawsandregs/hydrogensulfide/indexf.html.

- **TRI Reporting for Facilities Located in Indian country and Clarification of Additional Opportunities Available to Tribal Government under the TRI Program**

  EPA finalized a rule on April 19, 2012 (77 FR 23409) which requires each facility located in Indian country to submit TRI reports to EPA and the appropriate tribe, rather than to the state in which the facility is geographically located. The final rule also provides the tribal chairperson or equivalent elected official of a tribe with the same opportunities as the governor of a state with regard to TRI-related requests and petitions. The rule requirements are applicable for reporting year 2012 for TRI reports due to the Agency by July 1, 2013.

- **Pollution Prevention**

  In order to promote pollution prevention, EPA has increased the prominence and accessibility of the pollution prevention information reported in Sections 8.10 and 8.11 of the Form R. For example, the 2011 TRI National Analysis highlighted the parent companies that reported the greatest number of source reduction activities. To learn more, visit www.epa.gov/TRI/P2.

- **New Green Chemistry Source Reduction Codes**

  New source reduction codes that describe green chemistry practices have been added to the list of selections available for completing Section 8.10.

- **New in TRI-MEweb for RY 2012**

  o **New Certification Process in TRI-MEweb for RY 2012 Reporting**

    EPA has developed a new certification component within the TRI-MEweb application that will allow a facility to prepare any reporting year TRI Form R/A and transition directly into the certification process without leaving the TRI-MEweb application.

  o **New Real-Time Electronic Signature Agreement (ESA) Approval Option for New Certifying Officials**

    For this reporting year, EPA has implemented an alternative method for certifying officials to apply for and process an ESA in real-time using a third-party identity verification vendor named LexisNexis. Upon registering for a new CDX account at https://cdx.epa.gov, all newly appointed certifying officials will be prompted to consider using this third-party verification and authentication service to obtain their ESA approval in real-time instead of processing the traditional paper ESA form for EPA approval.

    An ESA form is a statement that declares that the registrant understands that any electronic signature executed with the electronic signature device is as legally binding as a handwritten signature and is required by EPA before any certifying official can certify and submit a TRI form created in TRI-MEweb.

  o **TRI-MEweb Pollution Prevention Upgrades**

    EPA has made several changes to TRI-MEweb to facilitate completion of the Pollution Prevention section of the Form R (Section 8). For example, TRI-MEweb now provides a Production Ratio/Activity Index Wizard to automatically calculate your Production Ratio/Activity Index (Section 8.9). TRI-MEweb also includes new tips and guidance for completing Sections 8.9-8.11 and allows the user to submit additional information about specific entries in Sections 8.9-8.10 if desired. (This information is then included in Sections 8.11 or 9.1 as appropriate.)

Other Important Information for Reporting Year 2012

EPA's Audit Policy. If you discover your facility is or may have been in violation of Section 313 of EPCRA (TRI Reporting), please refer to EPA's Policy entitled, "Incentives for Self-Policing: Discovery, Disclosure, Correction, and Prevention of Violations" (Audit Policy), 65 FR 19618, April 11, 2000. You may qualify for having all gravity-based penalties waived if your facility meets all nine (9) conditions of the Audit Policy. For more information on EPA's Audit Policy, see the Agency's website: http://www.epa.gov/compliance/incentives/auditing/auditpolicy.html.

EPA Enforcement Response Policy for TRI Revisions. On September 26, 1991, EPA published in the Federal Register, a "Notice Regarding Revisions to Toxic Chemical Release Inventory Reporting Forms under Section 313 of the Emergency Planning and Community Right-to-Know Act of 1986" (56 FR 48795-03). Section V of the Notice refers to the Agency's Enforcement and Penalties policy regarding Form R errors:

Facilities are reminded that there is a legal obligation to file an accurate and complete Form R report for each chemical by July 1 each year. EPA may take enforcement action and assess civil administrative penalties regarding corrections to errors in Form R reports that are not changes based on previously unavailable information or procedures which improve the accuracy of the data initially reported. The kinds of errors which may result in enforcement and in penalties include but are not limited to the following: (1) Errors caused by not using the most readily available information, for example, not using monitoring data collected for compliance with other regulations in calculating releases; (2) omitting a major source of emissions; (3) a mathematical or transcription or typographical error which seriously compromises the accuracy of

the information, and; (4) other errors which seriously affect the utility of the data, particularly errors in release reporting for which the facility has no records showing the derivation of the release calculation, and cannot provide a sufficient explanation of the report.

EPA's Small Business Compliance Policy. If you have 100 or fewer employees and discover that your facility is or may have been in violation of Section 313 of EPCRA (TRI Reporting), please refer to EPA's Small Business Compliance Policy. EPA will eliminate or significantly reduce penalties for small businesses that meet the conditions of the Policy, including voluntarily discovering violations and promptly disclosing and correcting them. This Policy implements Section 223 of the Small Business Regulatory Enforcement Fairness Act (SBREFA) of 1996. For more information, see the Agency's website: http://www.epa.gov/compliance/incentives/smallbusiness/index.html.

Parent Company Information. In past years, the Agency found that many facilities report inaccurate parent companies and/or Dun and Bradstreet numbers in Sections 4 and 5 of the TRI reporting forms. All facilities should verify the accuracy of facility and parent company information (e.g., DUNS number, parent company name). Related questions and answers are provided in Appendix I. *Please note that beginning in Reporting Year 2009, EPA started to pre-load standardized parent company names into the TRI-MEweb software that were researched from the prior year submission. This step was taken to improve the accuracy of parent company names as well as create a standard format for the names themselves. For example, only capital letters are used and all periods are eliminated from the parent names. In addition, standardized abbreviations are now used for common terms found in parent names such as 'CO for Company' and 'INC for Incorporated.' More detailed instructions appear in the Parent Company name sections of TRI-MEweb.*

For facilities still using paper TRI forms to report, a detailed listing of Parent Company names is provided at: http://www.epa.gov/tri/report/index.htm.

A. To verify the accuracy of your facility and parent company Dun and Bradstreet number and name, as required in Section 5 of both Form R and Form A, go to: https://www.dnb.com/product/dlw/form_cc4.htm or call 1-888-814-1435 to verify your information. Callers to the toll free phone number should understand that the Dun and Bradstreet support representatives will need to verify that callers requesting the DUNS numbers are agents of the business. Dun and Bradstreet recommends knowing basic information such as when the business originated, officer names, and the name, address, and phone number for the facility.

B. Facilities reporting to TRI should also make sure they are providing the parent company name and Dun and Bradstreet number as of December 31st of the current reporting year.

TRI-MEweb Reporting Year 2012 version

The TRI-MEweb application helps facilities fulfill their Emergency Planning and Community Right-to-Know (EPCRA) Section 313 and Pollution Prevention Act (PPA) Section 6607 obligations. TRI-MEweb is an interactive, intelligent, user-friendly web-based application tool that guides facilities through the TRI reporting experience. By leading prospective reporters through a series of logically ordered questions, TRI-MEweb streamlines the analysis needed to determine if a user must complete a Form R Report or if they meet the thresholds that allow them to use the Form A Certification Statement for a particular chemical.

The TRI-MEweb software provides the user with guidance for each data element on the reporting Form R and Form A Certification Statement. TRI-MEweb will check the data on the form for common errors and then prepare it for electronic transmission to be certified in CDX (see Figure 1 on page 2 for a flow diagram of the TRI-MEweb reporting process). All of the information contained in this RY 2012 Reporting Forms and Instructions manual is contained within the TRI-MEweb application.

EPA strongly recommends the use of TRI-MEweb to submit information to the TRI program. However, if you plan to submit hard copy forms for current year and prior year forms (RY 2011 and RY 2012), EPA provides *electronically fillable TRI forms* at: http://www.epa.gov/tri/reporting_materials/forms/ that can be completed using a computer and then printed, signed, and mailed.

EPA will accept only TRI-MEweb or paper submissions for RY 2012. The following are reasons why you should use TRI-MEweb in RY 2012:

• TRI-MEweb will guide reporting facilities through the RY 2012 changes to the reporting requirements described earlier in this section.

• A Web-enabled threshold determination tool is available on the TRI website at: http://www.epa.gov/tri/reporting_materials/threshold/ to help facilities determine whether they need to report to TRI. The threshold tool is useful in determining whether a facility meets all three threshold indices: number of employees, covered industry sector and chemical-specific thresholds.

- TRI-MEweb can be used to revise prior or current year data (RY 2005-RY 2012). TRI-MEweb stores data from the previous seven reporting years for each facility so that users no longer have to load the data each year.

  o If your facility needs to revise forms prior to RY 2005, EPA only accepts those revisions on hard-copy TRI forms.

- Facilities that reside in a state or in Indian Country participating in the TRI Data Exchange will have their TRI forms sent simultaneously to EPA and their state TRI representative via the Environmental Information Exchange Network.

  o For states or tribes <u>not</u> on the TRI Data Exchange (TDX), TRI-MEweb allows facilities to generate paper/disk/CD submissions for their state or tribe. Please note, however, that EPA's Data Processing Center does not accept disk/CD submissions.

- Facilities located within Indian country must find the corresponding Bureau of Indian Affairs (BIA) three-digit code for the appropriate tribe in Table V and fill in the code in the tribal name field in Section 4.1 of Form R or Form A or use the drop-down menu in TRI-MEweb to find the BIA code. <u>Currently, there are no tribes participating in EPA's TRI Data Exchange. Therefore, hard-copies of TRI forms must be mailed to the appropriate tribal point of contact.</u> Facilities using TRI-MEweb to fulfill their federal and tribal reporting requirements will be able to print a copy of the TRI form to be mailed to the appropriate tribal point of contact. (See Appendix E for link to points of contact and addresses.)

Uncertified TRI-MEweb Submissions. For Form R and/or Form A submissions via TRI-MEweb, a facility's registered certifying official must electronically sign the submission before it can be processed and entered into the TRI database. Uncertified electronic submissions created in TRI-MEweb are not considered to have completed the reporting requirement under EPCRA Section 313. Lack of certification will also prevent the submission from being processed by EPA.

How to Begin Using the Reporting Year 2012 TRI-MEweb Reporting Tool

Getting started. In early 2013, all technical contacts, preparers, and certifying officials from reporting facilities that filed TRI reports in the prior reporting year will be emailed their 6-digit alpha numeric *access key* for each facility account in TRI-MEweb. This unique access key is used to load any TRI data received by EPA from the past seven reporting years into its corresponding TRI-MEweb facility account. Facilities that have not received their unique 6-digit access key by February should call the Central Data Exchange (CDX) helpdesk (888) 890-1995. TRI-MEweb users may also use the access key that EPA sent to facility contacts in previous years, however, in some exceptional cases, the key may have been changed by EPA's Data Processing Center.

TRI-MEweb is accessed through EPA's Central Data Exchange (CDX). The TRI-MEweb application uses EPA's CDX network to transmit and certify electronic submissions to EPA. CDX allows facilities to submit a paperless report and receive instant receipt confirmation of their submission via the Internet. In addition, facilities that reside in a state or tribe participating in the TRI Data Exchange will have their forms sent simultaneously to EPA and their state or tribal TRI representative in electronic format.

How to Log into the Reporting Year 2012 TRI-MEweb Tool

Log in to your CDX user account to open the TRI-MEweb application. Preparers and certifying officials must have a CDX user account. During the CDX registration process, both will need to add the TRI-MEweb application to the CDX user account before a user can start preparing/certifying their TRI forms. Your web browser must have a security setting of TLS 1.0 enabled. Otherwise, the CDX login web page will appear as if it is broken. For detailed instructions for changing the security settings on your Web browser, go to the TRI-MEweb Resource Web page:
http://www.epa.gov/tri/reporting_materials/trimeweb/.

- If you cannot reset the password to your CDX account or have forgotten your CDX login name, please call the CDX Hotline (888) 890-1995.

- For assistance setting up a new CDX user account, adding your TRI role, adding the TRI-MEweb application to your CDX user profile, or for information on how to use the TRI-MEweb reporting tool, please visit: http://www.epa.gov/tri/reporting_materials/trimeweb/.

- CDX login can be accessed at: https://cdx.epa.gov/.

Two user roles involved in TRI Reporting. If your facility is using TRI-MEweb to prepare, transmit, and certify TRI forms to EPA, the application requires that two user roles be created. First, individuals who use TRI-MEweb to prepare and transmit data to CDX ("preparers") must have a role, and second, individuals who will certify pending TRI forms ("certifying officials") must have a role. Both preparers and certifiers must have CDX user accounts. TRI-MEweb allows the certifying official to also prepare TRI forms (dual role). All newly designated certifying officials with CDX accounts are required to submit an Electronic Signature Agreement (ESA) application to EPA for approval before facilities can certify TRI forms submitted to EPA and their appropriate state or tribe. (See Section A.2 for details on ESA processing)

TRI-MEweb can import previous year data into current year chemical forms. TRI-MEweb now has the ability to import prior year data (if RY 2011 data was provided in the previous year by reporting facility) into each selected current year TRI chemical form. Importing data is optional; however, it can accelerate data entry if the same chemicals are reported to EPA every reporting year.

Easier certification process now built into TRI-MEweb. EPA has developed a new certification process within TRI-MEweb that allows a facility to prepare any reporting year TRI Form R or Form A Certification Statement and go directly into the certification process. In past years, certifying officials had to leave TRI-MEweb to do the certification process. A user with a dual role (preparer role and certifying official role) will no longer need to log out of the system to certify the forms prepared in the TRI-MEweb application. The certifying official will simply login into their CDX account and open TRI-ME-web to certify any TRI forms with certification pending.

1. **New security requirements.** All certifying officials responsible for an ESA approved in a prior year must comply with new security

standards by choosing and answering 5 questions from a list of 20. One of the answered questions will be used to verify and authenticate the identity of the certifying official as part of the electronic signature process. As part of the new process, you will also be required to provide the TRIFID of your reporting facility and electronically sign a TRIFID Certification Agreement to obtain access to any pending submissions. The new certification module in TRI-MEweb captures and stores this information for all certifying officials. A user with a dual role (preparer role and certifying official role) will no longer need to logout to begin certifying forms prepared in the TRI-MEweb application.

2. **Electronically sign your current Electronic Signature Agreement (ESA).** If you have an existing ESA, you must "electronically sign" that ESA to ensure that it continues to be valid under the new electronic reporting requirements. From now on, revalidated ESAs will be referred to as CDX ESAs. All new certifying officials must sign a CDX ESA in TRI-MEweb to allow for expanded reporting functionality. Certifiers are not required to mail a hardcopy of this expanded CDX ESA to the EPA's Data Processing Center for approval. Certifiers completing this step will be able to add facility account(s) without having to generate a new hardcopy ESA.

3. **Add the TRI Facility Identification Number (TRIFID) to the new certification module in TRI-MEweb.** If you are a certifying official with a current ESA, who is logging into a CDX user account for the first time, you will be prompted to proceed directly to the "Manage TRIFIDs" section under the "Certify" tab. Once there, you must add all TRIFIDs for which TRI forms will be transmitted. If any TRI forms have been previously sent, they will appear under the "**Pending Submission**" subtab.

New "real-time" Electronic Signature Agreement (ESA) approval option for new certifying officials. EPA now provides an alternative method for certifying officials to process ESAs in real-time using a third-party identity verification vendor named LexisNexis. In the past, all new ESA applications were required to be mailed in for approval, a process that took about two weeks. Now all new certifying officials will be prompted to consider using LexisNexis to obtain their ESA approval upon registering for a new CDX account at https://cdx.epa.gov. ESAs approved using LexisNexis will receive electronic notification in seconds, or "real time".

Another significant advantage of the LexisNexis method, besides obtaining immediate ESA approval, is that the LexisNexis approval is applicable to multiple CDX system flows (TRI). Programs like eTSCA and Risk Management Plan (RMP eSubmit) will be able to share the security credentials offered by the CDX ESA obtained under TRI. To obtain this real-time approval, the certifying official must provide personal identity authentication information such as name, address, etc. Please note that EPA does not collect any personal information from our users. The use of these third party verification and identification widgets is common in banking systems.

However, for those new certifying officials that do not wish to provide personal information to a third-party vendor, or who fail the LexisNexis method for some reason, the traditional paper ESA form will still be available. **The hard-copy ESA approval process requires the printing, completion, and mailing of an electronic signature agreement form (see page 4 for where to send form). Please allow adequate time for the mailing and processing of this form, which is estimated to take a minimum of five (5) business days.** Facilities that do not have a signed ESA, either electronically or on hardcopy, will not be able to certify forms in TRI-MEweb. It is suggested that certifying officials complete their ESA well in advance of the reporting deadline.

TRI-MEweb can be used to submit revisions or withdrawals for data reported between RY 2005 through RY 2012. If you need to revise or withdraw prior or current year data, we encourage you to submit revisions using the TRI-MEweb application. If you need to revise or withdraw forms prior to RY 2005, you must provide these requests on hardcopy TRI forms.

Facilities are reminded that there is a legal obligation to file an accurate and complete Form R report for each chemical by July 1 each year. EPA may take enforcement action and assess civil administrative penalties regarding corrections to errors in Form R reports that are not changes based on previously unavailable information or procedures which improve the accuracy of the data initially reported. The kinds of errors which may result in enforcement and in penalties include but are not

limited to the following: (1) Errors caused by not using the most readily available information, for example, not using monitoring data collected for compliance with other regulations in calculating releases; (2) omitting a major source of emissions; (3) a mathematical or transcription or typographical error which seriously compromises the accuracy of the information, and; (4) other errors which seriously affect the utility of the data, particularly errors in release reporting for which the facility has no records showing the derivation of the release calculation, and cannot provide a sufficient explanation of the report.

Electronic Facility Data Profile (eFDP). Reporting facilities may confirm and review their submitted TRI data to EPA by viewing their electronic Facility Data Profile (eFDP) on the Internet by logging into their CDX account and clicking the *TRI-MEweb: TRI Made Easy* link from their *MyCDX* page. This will open the "Welcome" page of the TRI-MEweb application. On the "Welcome" page, follow the instructions for viewing the eFDP. If the Technical Contact provides an email address in the Form R/Form A Certification Statement, they will receive an email notifying them when their eFDP has been updated and published for review in TRI-MEweb.

TRI-MEweb User Resources

- TRI-MEweb website: http://www.epa.gov/tri/reporting_materials/trimeweb/. Any service notification that may impact your electronic reporting will be posted on this webpage. This webpage also contains links to reference materials that will assist you in fulfilling your reporting requirements.

- TRI-MEweb Online tutorials: http://www.epa.gov/tri/reporting_materials/tutorials/tutorial_index.html. If you need additional assistance on how to perform the different tasks within the Web-based application reporting tool, we advise you to view the TRI-MEweb Online Tutorials These tutorials provide step-by step instruction on how to use TRI-MEweb.

- TRI Information Center Hotline [(800) 424-9346 - select option 3] and CDX Help Desk (888) 890-1995: These hotlines provide CDX/TRI-MEweb technical support and reporting requirements assistance to help reporting facilities meet their TRI reporting obligations.

A. General Information

Reporting to the Toxic Chemical Release Inventory (i.e., Toxics Release Inventory (TRI)) is required by Section 313 of the Emergency Planning and Community Right to Know Act (EPCRA, or Title III of the Superfund Amendments and Reauthorization Act of 1986), Public Law 99 499. The information contained in the Form R constitutes a "report," and the submission of a report to the appropriate authorities constitutes "reporting."

The Pollution Prevention Act, of October, 1990 (Pub. L. 101 508), added reporting requirements to the Form R. These requirements affect all facilities required to submit a Form R under Section 313 of EPCRA. The data were required beginning with reports for calendar year 1991.

Reporting is required to provide information to the public on releases and other waste management of EPCRA Section 313 chemicals in their communities and to provide EPA with release and other waste management information to assist the Agency in determining the need for future regulations. Facilities must report the quantities of routine and accidental releases, and releases resulting from catastrophic or other onetime events of EPCRA Section 313 chemicals, as well as the maximum amount of the EPCRA Section 313 chemical on-site during the calendar year and the amount contained in wastes managed on-site or transferred off-site.

A completed Form R or Form A must be submitted for each EPCRA Section 313 chemical manufactured, processed, or otherwise used at each covered facility as described in the reporting rules in 40 Code of Federal Regulations (CFR) Part 372 (originally published February 16, 1988, in the *Federal Register* and November 30, 1994, in the *Federal Register* (for Form A)).

A.1 Who Must Report

- Section 313 of EPCRA requires that reports be filed by owners and operators of facilities that meet all of the following criteria.

- The facility has 10 or more full-time employee equivalents (i.e., a total of 20,000 hours or greater; see 40 CFR 372.3); and

- The facility is included in a North American Industry Classification System (NAICS) code listed in Table I. NAICS codes found in Table I

correspond to the following Standard Industrial Classification (SIC) Codes: SIC 10 (except 1011, 1081, and 1094), 12 (except 1241), 20-39, 4911 (limited to facilities that combust coal and/or oil for the purpose of generating electricity for distribution in commerce), 4931 (limited to facilities that combust coal and/or oil for the purpose of generating electricity for distribution in commerce), 4939 (limited to facilities that combust coal and/or oil for the purpose of generating electricity for distribution in commerce), 4953 (limited to facilities regulated under RCRA Subtitle C, 42 U.S.C. Section 6921 *et seq.*), 5169, 5171, and 7389 (limited to facilities primarily engaged in solvents recovery services on a contract or fee basis); and

- The facility manufactures (defined to include importing), processes, or otherwise uses any EPCRA Section 313 chemical in quantities greater than the established threshold in the course of a calendar year.

Executive Order 13423 extends these reporting requirements to federal facilities, regardless of their SIC or NAICS code.

A.2 How to Submit Forms

Facilities can use TRI-MEweb or paper for submitting Form R(s) and/or Form A Certification Statement(s).

A.2.a How to Submit Form R(s) and/or Form A(s) Certification Statement Electronically to EPA via the Central Data Exchange (Using the TRI-MEweb Application)

The preferred method to report your toxic chemical release data to your federal, state or tribal TRI authorities is by the use of the web-based TRI-MEweb application via EPA's Environment Information Exchange Network (EIEN). The EIEN is a partnership among state, tribes, territories, and EPA to deliver critical environmental information. The Central Data Exchange (CDX) is the point of entry for all TRI reporting facilities to the EIEN for environmental data submissions to all EIEN partners. CDX also hosts the TRI-MEweb reporting tool software. TRI-MEweb allows facilities to file a paperless report through the Internet, significantly reducing data errors, and facilities receive instant receipt confirmation of their submission. There are several other advantages to using TRI-MEweb for TRI reporting: prior year TRI form data are imported into current year forms to expedite reporting, allowing a certifier to submit an electronic signature that allows one to file paperless forms, and error checking software that ensures higher data quality.

Preparers that use TRI-MEweb to transmit forms through the Internet via CDX and reside in a state or tribe participating in the TRI Data Exchange (TDX) will have their forms sent simultaneously to EPA and their respective state or tribal officials via the CDX (Internet submissions are not available for trade secret claims). Once a TRI submission is certified, it will be electronically forwarded to state or tribal officials and your obligation to report to EPA and your state or tribe will be satisfied.

Please be aware that if your facility does not reside in a state or tribe participating in the TDX, just transmitting TRI forms via the Internet does not satisfy your state or tribal reporting requirements for your facility. You must report to your state or tribe

separately and in the required format specified by your state or tribe. However, TRI-MEweb can still be used by the reporting facility to prepare and print the proper paper TRI forms for your state or tribal submission that is not in the TDX. Certifying officials must sign any TRI forms printed from TRI-MEweb and physically mail these completed paper TRI forms to non-TDX states or tribes. Do not send forms in draft format, or printed copies from the TRI-MEweb application to EPA's Data Processing Center.

If you choose to submit your federal submission using TRI-MEweb via the Internet, do not send duplicate paper copies of the reports to EPA's Data Processing Center (DPC).

Electronic Signature Agreement

There are two user roles involved in the reporting process to EPA of your TRI data; a preparer role and a certifying official role. Figure 1 illustrates how these two roles are involved in the TRI reporting process. The "Preparer" is the person who will be preparing the TRI forms for submission in TRI-MEweb but is not authorized to certify them. The "Certifying Official" is the person

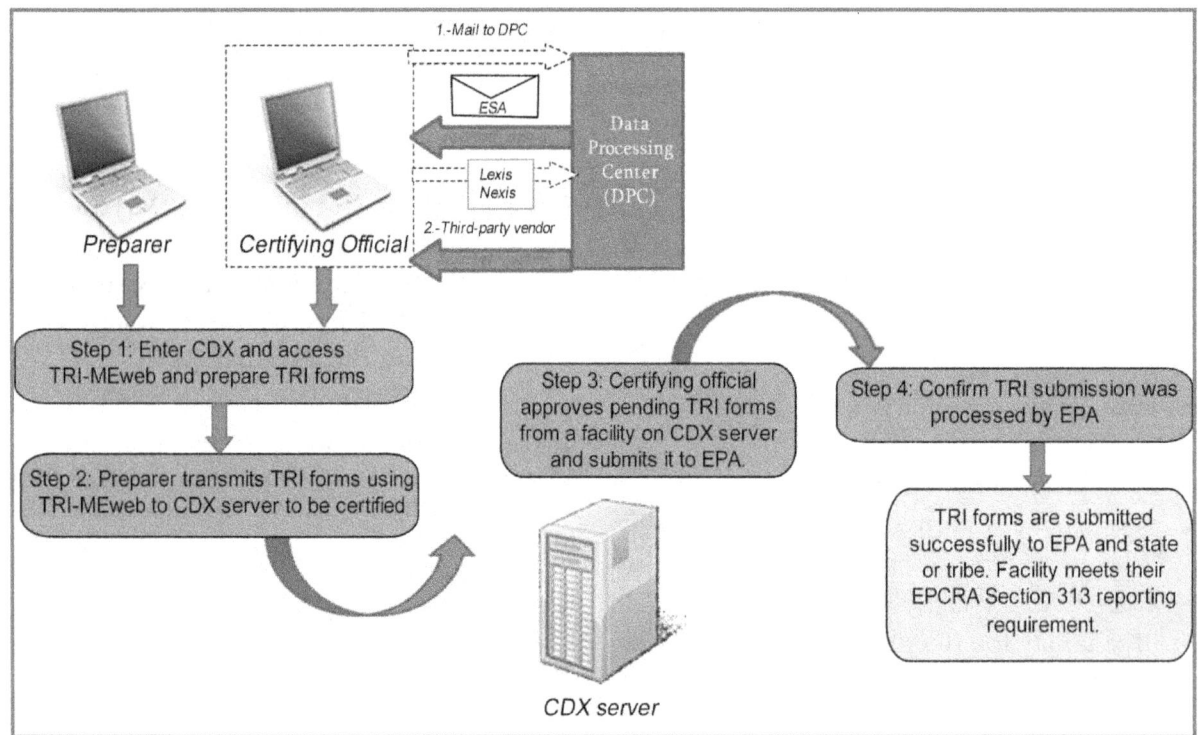

Figure 1. TRI-MEweb's Preparation, Transmission, Certification and Submission Steps

of authority at a facility or a legal representative of the facility that will be certifying the data contained in the transmitted TRI Form R or Form A Certification Statement in TRI-MEweb to EPA and their state or tribe. Certifying officials may also prepare forms (dual role). However, the preparer role cannot certify TRI forms that have been transmitted to CDX. Both of these TRI roles require creating/having a CDX user account and adding the TRI-MEweb application to their MyCDX profile. Step-by-step instructions on how to create your CDX user account for your new certifying official can be found on the TRI-MEweb Resource web page: http://www.epa.gov/tri/reporting_materials/trimeweb/.

An Electronic Signature Agreement (ESA) form is a statement that declares that the registrant understands that any electronic signature executed with the electronic signature device is as legally binding as a handwritten signature and is required by EPA before any certifying official can certify and submit a TRI form created in TRI-MEweb. An ESA request is needed in RY 2012 for certifying officials that have been newly designated for a facility. In RY 2012, all existing certifying officials will be required to electronically sign a one-time update to their ESA record (see next section for more information). ESAs are created upon the certifying official creating a new CDX user account and applying for a certifying official role. In RY 2012, there are two options available to obtain an ESA approval from EPA.

Option 1 - LexisNexis real-time ESA approval. EPA now provides an alternative method for certifying officials to process ESAs in real-time using a third-party identity verification vendor named LexisNexis. This real-time approval is possible because personal identifying information is provided voluntarily by the certifying official to a third-party vendor (EPA does not collect any personal information from our users) to authenticate their identity. Third party verification and identification widgets are commonly used in the banking system. The most significant benefit from the LexisNexis method is that users will no longer need to wait up to 5 business days for an ESA approval by EPA. This alternate ESA approval method is optional. If the certifying official does not wish to provide personal information to a third-party vendor; they should

submit a paper ESA form instead well in advance ahead of the July 1 deadline.

Option 2 - Paper ESA form. A printable ESA form can be generated during the CDX registration process. The ESA form must be signed and mailed to EPA's Data Processing Center (DPC in figure 1) for approval before the certifying official can begin to certify any TRI forms transmitted by the preparer to CDX using TRI-MEweb. Hardcopy ESA approval may take up to five business days, so please plan accordingly or consider the option one, LexisNexis. Multiple TRIFIDs can also be added to a single hardcopy ESA form. All newly assigned TRIFIDs will be listed in the printout of the ESA document. TRI-MEweb is updated when the ESA is approved.

All existing certifying officials in RY 2012. All existing certifying officials are required to electronically sign an update to their CDX ESA to allow for electronic reporting. No hardcopy of this CDX ESA form is required to be mailed to EPA's Data Processing Center for approval if a signed hardcopy ESA has been previously filed because the user has already had their identify verified with their originally submitted TRI ESA. This update is only required in RY 2012. The processing and activation of this CDX ESA is done in real-time and takes advantage of a security upgrade to the TRI reporting process. As a result of this upgrade, all existing certifying officials will complete the 20-5-1 Questionnaire upon logging into CDX for the first time during the RY 2012 reporting season. The 20-5-1 Questionnaire requires you to answer five personal knowledge questions (e.g., "Favorite Color," "Name of Pet," "Mother's Name," etc) from a list of twenty. Upon subsequent logins, one answer is required from the five questions previously answered to verify user identity. The new ESA for electronic reporting will provide the benefit to all existing certifying officials of being able to add facility account(s) without having to generate a new hardcopy ESA to be printed, signed and mailed to EPA for approval before the certifying official can certify any pending forms from the newly added reporting facility. Certifying officials will only need to add the TRIFID of the new facility account and their 6-digit access key on the "Manage TRIFIDs" page under the "Certify" tab in TRI-MEweb to access the history of any pending submission and taken action to either cancel or certify any transmitted chemical form(s). They will also need to sign a TRIFID Signature Agreement for any additional TRIFIDs added to the TRI-MEweb application

Accidental deletion of ESA in TRI-MEweb. The TRI-MEweb application also has the capability to manage user profiles (previously authorized preparers or certifying officials) that have been granted access to facility accounts. This capability includes revoking approved ESA(s) for any certifying official(s) that have left the facility's payroll or is no longer authorized to certify forms. An ESA could also be accidently revoked by the preparer. If this occurs, there is a 45-day grace period to get the ESA reactivated by the CDX helpdesk without having to send a paper form to EPA for re-approval. An email notification is sent to the affected certifying official by CDX when any ESA has been revoked within TRI-MEweb.

Mailing information for EPA's Data Processing Center can be found below or on the TRI website under Contact Us: http://www.epa.gov/tri/index.htm.

Where to send hardcopy ESA forms

Send ESA approval requests by *regular mail* to the following address:

> Attention: TRI ESA Approval Request
> TRI Reporting Center
> P.O. Box 10163
> Fairfax, VA 22038

Send ESA approval requests by *certified mail or overnight mail* to the following address:

> Attention: TRI ESA Approval Request
> CGI Federal, Inc
> c/o EPA Reporting Center
> 12601 Fair Lakes Circle
> Fairfax, VA 22033

For questions or additional information about CDX, please see: http://www.epa.gov/cdx.

A.2.b   How to Submit Paper Form R (s) and/or Form A Certification Statement(s)

It is EPA's ultimate goal to move away from processing paper submissions and to receive all TRI submissions via CDX. Although EPA strongly discourages paper submissions due to increased possibility of errors, paper submissions are currently still accepted. Paper submissions must be sent to both EPA and the state or the designated official of an Indian tribe. If a report is not received by both EPA and the state (or the designated official of an Indian tribe), the submitter is considered out of compliance and subject to enforcement action. Facilities submitting paper forms must use the corresponding reporting year forms. To facilitate the completion and processing of paper forms, EPA is providing electronically fillable reporting forms that can be completed prior to printing for RY 2011 and RY 2012 TRI forms. Most paper TRI forms are submitted and received by EPA for the current reporting year and the prior year. EPA strongly encourages facilities to use this new tool to complete forms prior to printing them. The fillable reporting forms can be found on URL: http://www.epa.gov/tri/reporting_materials/index.html.

Send paper forms by regular mail to:
> TRI Reporting Center
> P.O. Box 10163
> Fairfax, VA 22038

Send paper forms by certified mail or overnight mail (i.e. Fed Ex, UPS, etc.) to:
> CGI Federal, Inc
> c/o EPA Reporting Center
> 12601 Fair Lakes Circle
> Fairfax, VA 22033

E-mailed submissions will not be accepted.

State and Tribal Submissions. If the facility is in a state that is not in EPA's TRI Data Exchange (TDX) system, then the facility must also send a copy of the report to the state. To verify if your state is or is not in the TDX system, go to: http://www.epa.gov/tri/stakeholders/state/state_exchange/index.htm. "State" also includes: the District of Columbia, the Commonwealth of Puerto Rico, Guam, American Samoa, Marshall Islands, the U.S. Virgin Islands, the Commonwealth of the Northern Mariana Islands, and any other jurisdiction and Indian country. Refer to Appendix E for the appropriate state submission addresses.

Facilities that are located within a tribe's Indian country will need to find their three-digit Bureau of

Indian Affairs (BIA) tribal code for their Indian country name in Table V and fill in the code in the "City/County/Tribe/State/ZIP code" field on the Form R or Form A in Section 4.1. A hard-copy of TRI forms must be mailed to the Indian country's Chief Executive Officer because most Indian country entities are not members of TDX in RY 2012. Some tribes have entered into a cooperative agreement with states; in this case, report submissions should be sent to the entity designated in the cooperative agreement. Facilities using TRI-MEweb to fulfill their federal and tribal reporting requirements under EPCRA Section 313 will be able to print a hard-copy of the TRI form to be mailed to their Indian country's Chief Executive Officer.

A.3   Trade Secret Claims

For any EPCRA Section 313 chemical whose identity is claimed as trade secret, you must submit to EPA two versions of the substantiation form as prescribed in 40 CFR Part 350, published July 29, 1988, in the *Federal Register* (53 FR 28772) as well as two versions of the EPCRA Section 313 report. The current substantiation form is available on the TRI website at:
http://www.epa.gov/tri/reporting_materials/index.html. One set of reports, the unsanitized version, must provide the actual identity of the EPCRA Section 313 chemical. The other set of reports, i.e., the "sanitized" version, must provide a generic class or category for the chemical that is structurally descriptive of the EPCRA Section 313 chemical. If EPA deems the trade secret substantiation form valid, only the sanitized set of forms will be made available to the public.

Further explanation of the trade secret provisions is provided in Part I, Sections 2.1 and 2.2, and Part II, Section 1.3, of the instructions.

In summary, a complete report to EPA for an EPCRA Section 313 chemical claimed as a trade secret must include all of the following:

- A completed unsanitized version of Form R or Form A report including the EPCRA Section 313 chemical identity (staple the pages together); and
- A sanitized version of a completed Form R or Form A report in which the EPCRA Section 313 chemical identity items (Part II, Sections 1.1

and 1.2) have been left blank but in which a generic chemical name that is structurally descriptive has been supplied (Part II, Section 1.3) (staple the pages together); and
- A completed unsanitized version of a trade secret substantiation form (staple the pages together); and
- A sanitized version of a completed trade secret substantiation form (staple the pages together).

Securely fasten all four reports together.

Some states or tribes also require submission of both sanitized and unsanitized reports for EPCRA Section 313 chemicals whose identity is claimed as a trade secret. Others require only a sanitized version. Facilities may jeopardize the trade secret status of an EPCRA Section 313 chemical by submitting an unsanitized version of the EPCRA Section 313 report to a state agency or Indian tribe that does not require unsanitized forms. You may identify an individual state or tribe's submission requirements by contacting the appropriate state or tribe designated EPCRA Section 313 contact (see Appendix E).

Where to send your trade secret submission

Please send only trade secret submissions to the P.O. Box below. Send trade secret submissions by *regular mail* to:
Attention: EPCRA Substantiation Packages
TRI Reporting Center
P.O. Box 10163
Fairfax, VA 22038

Send trade secret submissions by *certified mail or overnight mail* (i.e. Fed Ex, UPS, etc.) to:
Attention: EPCRA Substantiation Packages
CGI Federal, Inc.
c/o EPA Reporting Center
12601 Fair Lakes Circle
Fairfax, VA 22033

A.4   Recordkeeping

Sound recordkeeping practices are essential for accurate and efficient TRI reporting. It is in the facility's interest, as well as EPA's, to maintain records properly. Facilities must keep a copy of each report filed for at least three years from the date of submission. These reports will be of use when completing future reports.

Facilities must also maintain those documents, calculations, worksheets, and other forms upon which they relied to gather information for prior reports. In the event of a problem with data elements on a facility's Form R or Form A report, EPA may request documentation from the facility that supports the information reported.

EPA may conduct data quality reviews of Form R or Form A submissions. An essential component of this process involves reviewing a facility's records for accuracy and completeness. EPA recommends that facilities keep a record for those EPCRA Section 313 chemicals for which they did not file EPCRA Section 313 reports.

EPA also recommends keeping records of all documentation containing your CDX account information for your preparer(s) and certifying official(s) that use TRI-MEweb to prepare and certify the reporting facility's TRI Form R and/or Form A. These CDX documents include the Electronic Signature Agreement (ESA) and the facility's unique 6-digit alphanumeric access key.

Records to maintain include:

- Previous years' EPCRA Section 313 reports;
- EPCRA Section 313 Reporting Threshold Worksheets;
- Engineering calculations and other notes;
- Purchase records from suppliers;
- Inventory data;
- EPA (NPDES) permits and monitoring reports;
- EPCRA Section 312 Tier II Reports;
- Monitoring records;
- Flowmeter data;
- RCRA Hazardous Waste Generator's Report;
- Pretreatment reports filed by the facility with the local government;
- Invoices from waste management companies;
- Manufacturer's estimates of treatment efficiencies;
- RCRA manifests;
- Process diagrams that indicate emissions and other releases;

- Records for those EPCRA Section 313 chemicals for which they did not file EPCRA Section 313 reports; and
- CDX account information including unique 6-digit access key to pre-load facility account into TRI-MEweb and copies of the Electronic Signature Agreement (s) submitted to EPA for approval.

A.5 How to Revise, Withdraw or Cancel TRI Data

A.5.1 Revising TRI Data

Facilities that filed a Form R and/or Form A Certification Statement under EPCRA Section 313 may submit a request to revise a form that was previously submitted, stored in EPA's historical database called the Toxics Release Inventory Processing System (TRIPS), and made available to the public through Envirofacts and TRI Explorer. Facilities may request a revision for one or more of the following reasons:

Revision codes:

- RR1 - New Monitoring Data
- RR2 - New Emission Factor(s)
- RR3 - New Chemical Concentration Data
- RR4 - Recalculation(s)
- RR5 - Other Reason(s)

Please note that late submissions for chemicals not reported in a previous reporting year are not considered revisions for that year.

Facilities are reminded that there is a legal obligation to file an accurate and complete Form R or Form A report for each chemical by July 1 each year. EPA may take enforcement action and assess civil administrative penalties regarding corrections to errors in Form R reports that are not changes based on previously unavailable information or procedures which improve the accuracy of the data initially reported. The kinds of errors which may result in enforcement and in penalties include but are not limited to the following: (1) Errors caused by not using the most readily available information, for example, not using monitoring data collected for compliance with other regulations in calculating releases; (2) omitting a major source of emissions; (3) a mathematical or transcription or typographical error which seriously compromises the accuracy of the information, and; (4) other errors which seriously affect the utility of the data, particularly errors in release reporting for which the facility has no records showing the derivation of the release calculation, and cannot provide a sufficient explanation of the report.

How do I revise my submission(s)?

If you have determined that your facility wishes to revise a TRI submission, you must send your revised report to EPA and the appropriate state or tribal agency. For submitting a revision to EPA, please use one of the following methods:

- EPA will accept revisions for Reporting Year 2005 through the current year via TRI-MEweb. This is EPA's preferred method to submit revisions.
- EPA will accept revisions prior to Reporting Year 2005 on paper forms using the TRI form applicable for the particular year being revised. Keep in mind that EPA does not key-in paper-form data into TRI's database until after the July 1 deadline.

1. TRI-MEweb. The preferred method for revising TRI forms from Reporting Year 2005 through the current year is to use TRI-MEweb. TRI-MEweb provides several advantages compared to hard-copy reporting, such as automated error-checking software, automatically updating Section 8 calculations when values are modified, and electronically confirming EPA's receipt of a submitted revision. All revisions created using TRI-MEweb are required to be transmitted to CDX and certified before being processed by EPA's DPC. If you have questions about using TRI-MEweb to revise your Form R/A, please refer to the TRI-MEweb *tutorial* page at: http://www.epa.gov/tri/reporting_materials/tutorials/tutorial_index.html.

2. Hard Copy Form. EPA strongly discourages paper submissions due to the increased possibility of data entry errors; however, if necessary, you may revise a previously submitted hard-copy form or revision by using either 1) a photocopy of the original TRI Form R or Form A submitted to US

EPA or 2) a blank Form R or Form A or 3) using your Facility Data Profile report.

- **Photocopy of Original Submission.** You may submit a photocopy of your original submission (from your records) with the corrections made in blue ink. Please re-sign and re-date the certification statement in Section 3 of the TRI Form R or Form A Certification Statement. For RY 2007 revisions and beyond, please enter the appropriate revision code(s) in the "Revision" box on the first page of the reporting form. You may enter up to two revision codes on the form. For RY 2006 and prior years, please enter an "X" in the space marked "Enter 'X' here if this is a revision," on Page 1 of the form.

- **Blank Form.** Hard copy submissions may be submitted using the TRI Form R and/or Form A applicable for that particular reporting year or the most recent form available. For revisions submitted by hard copy, EPA recommends the use of the most recent form that can be downloaded from the TRI website. You can also request prior year reporting forms under the *Contact Us* link on the TRI web site. For RY 2011 and current year forms, EPA recommends using the electronically fillable forms to submit your revision. RY 2010 and prior year forms are not electronically fillable and must be completed by hand. For RY 2007 revisions and beyond, please enter all information including the appropriate revision code(s). For RY 2006 and prior years, please enter all information including an "X" in the space marked "Enter 'X' here if this is a revision," on Page 1 of the form. Please sign and date the certification statement on Page 1.

Electronic Facility Data Profile (eFDP). eFDPs are made available by EPA's Data Processing Center (DPC) to a reporting facility in response to any submission EPA's DPC receives and has processed successfully into TRI's database. All reporting facilities may access and print their eFDP report via TRI-MEweb and make any needed corrections in blue ink to TRI data that have been submitted previously to EPA. These hard-copy eFDP forms with hand-written corrections must be mailed to EPA's DPC to be keyed into the TRI database. Please note that a certification statement on the bottom of the eFDP report that must be signed by a facility owner/operator or senior management official if using the eFDP hard-copy to make a revision. Please go to Appendix C for more information on the eFDP. Hard-copy eFDPs with hand-written corrections are still vulnerable to human error when manually keyed by the DPC into the TRI database. Therefore, EPA recommends that facilities use TRI-MEweb to send corrections in response to their eFDPs.

Mailing Address for Hard Copy Forms. Send revision requests by *regular mail* to the following address:

> Attention: TRI Revision Request
> TRI Reporting Center
> P.O. Box 10163
> Farifax, VA 22038

Send revision requests by *certified mail or overnight mail* to the following address:

> Attention: TRI Revision Request
> CGI Federal, Inc.
> c/o EPA Reporting Center
> 12601 Fair Lakes Circle
> Fairfax, VA 22033

A.5.2 Withdrawing TRI Data

Facilities that filed a Form R and/or Form A Certification Statement under EPCRA Section 313 may submit a request to withdraw a form that was previously submitted, stored in the Toxics Release Inventory Processing System (TRIPS), and made available to the public through Envirofacts and TRI Explorer. EPA may periodically review withdrawals. Facilities may request a withdrawal for one or several reasons, such as:

Withdrawal codes:

- WT1 - Did not meet the reporting threshold for manufacturing, processing, or otherwise use

- WT2 - Did not meet the reporting threshold for number of employees

- WT3 - Not in a covered NAICS Code

- WO1 - Other reason(s)

How do I withdraw my submission(s)?

If you have determined that your facility wishes to withdraw a TRI submission, you must send your request to EPA and the appropriate state agency. Keep in mind that successfully completed withdrawal requests will permanently delete the chemical release data provided by the reporting facility and processed into TRI's publicly available database.

If the reporting facility needs to make a correction to data submitted to EPA, a revision is easier to process by a reporting facility than withdrawing incorrect TRI forms and resubmitting them to EPA.

For submitting a withdrawal to EPA, please use one of the following methods:

TRI-MEweb. The preferred method for withdrawing TRI forms from Reporting Year 2005 through the current year is to use TRI-MEweb. Withdrawals can only be done for TRI submissions that have been properly transmitted, certified and processed by EPA. If you have questions about using TRI-MEweb to withdraw your Form R/A, please refer to the TRI-MEweb *tutorial* page at: http://www.epa.gov/tri/reporting_materials/tutorials/tutorial_index.html.

Hard Copy Form. All other withdrawal requests may be submitted by hard copy. The withdrawal code(s) should be entered in the "Withdrawal" box on the first page of the reporting form. You may enter up to two withdrawal codes on the form. There are two requirements for requesting a withdrawal on hardcopy, as follows:

- *Reporting Year 2007 Forward.* You may submit a photocopy of your original submission (from your records). Using blue ink, re-sign and re-date the certification statement on Page 1 and enter the appropriate withdrawal code(s) in the space provided on Page 1 of the form.

- *Reporting Year 2006 and Prior Years.* Please submit a photocopy of the form you wish to withdraw (from your file), and attach – as a cover page – Page 1 of the current year's reporting form, which includes a field for the withdrawal codes. Using blue ink, please sign and date the certification statement and enter all information including the appropriate withdrawal code(s) in the space provided on Page 1 of the current year's form.

Mailing Address for Hard Copy Forms. Send withdrawal requests by *regular mail* to the following address:

> Attention: TRI Withdrawal Request
> TRI Reporting Center
> P.O. Box 10163
> Fairfax, VA 22038

Send withdrawal requests by *certified mail or overnight mail* to the following address:

> Attention: TRI Withdrawal Request
> CGI Federal, Inc
> c/o EPA Reporting Center
> 12601 Fair Lakes Circle
> Fairfax, VA 22033

A.5.3 Canceling a TRI Submission

Cancelling a TRI form is not applicable to hard-copy TRI forms. However, different situations may require a TRI-MEweb user to cancel an electronic TRI submission. For instance, a facility's preparer or certifying official may determine that a draft electronic submission(s) requires cancellation because the facility's chemical release did not, in fact, meet the reporting thresholds of EPCRA Section 313.

Another reason why a TRI-MEweb submission may require cancellation is if a preparer has determined that a correction is needed on a TRI form that is pending certification in CDX, but has not yet been certified. In order to edit a TRI form in TRI-MEweb that is pending certification to CDX, the preparer will need to cancel the transmitted submission with a *Pending Certification* status in order to make the additional corrections in TRI-MEweb and retransmit the original submission or revision to CDX to be certified. EPA is considering issuing a Notice of Non-compliance for TRI Forms that have been transmitted to CDX but are not certified.

A preparer or a certifying official cannot cancel a TRI form submission that has already been transmitted and certified by the certifying official. If a chemical form has a status of *Certified and Sent to EPA* in TRI-MEweb it cannot be called back to be

edited or corrected. To change or remove data that has already been transmitted, certified and submitted to EPA to be processed, either revise or withdraw the submission.

How to Cancel a TRI Submission that has not been Certified.

If your facility decides not to complete the certification process for any pending electronic submission(s) transmitted to CDX by TRI-MEweb, you should **CANCEL** the submission(s) using one of the following methods:

By the Preparer: The preparer may use the TRI-MEweb application to cancel any unwanted pending submission(s). In TRI-MEweb, the preparer must click the "**Prepare**" tab, choose the Reporting Year corresponding to the unwanted submission(s) from the "**Select Year**" tab, choose the appropriate facility from the "**Select Facility**" tab, and select the chemical form to be cancelled from the **Select a Form** page. Next, the preparer must click the "**Review**" tab. Then, the preparer must locate the submission that includes the chemical form they wish to cancel and select its radio button from the *Pending Submission Summary Table* on the **Reporting Summary** page. Next, they must click the "**Cancel**" button and confirm the cancellation on the next page. Note: ALL chemical forms that were included in the selected submission will be canceled.

By the Certifying Official: The certifying official may also cancel any unwanted TRI submission(s) pending certification. The certifying official must log into their CDX account and click the "**TRI-MEweb: TRI Made Easy – Prepare/Certify Submission**" link from their MyCDX page. This will open the "**Welcome**" page of the TRI-MEweb application and then select the "**Certify**" tab. If certifying official does not find the TRIFID for their reporting facility with pending submissions listed, they gain access to that facility account by entering the access key on the "Enter Facility's Access Information" page and signing the TRIFID Certification Agreement on the "Manage TRIFIDs for Certification" page and clicking the "**Next**" button. The electronic signature widget will pop-up to confirm your authorized access to the facility account. Upon successfully authentication of user identity, you may begin the cancellation process on the "**Pending Submissions**" page under the "**Certify**" tab. You may view the content of the submission by clicking the "*View Submission*" icon to confirm that this is the correct submission to be cancelled. Select the "**Cancel**" radio button to cancel submission and select "**Next**" to confirm request. If you have questions about using TRI-MEweb to cancel your Form R or Form A Certification Statement submission, please refer to the TRI-MEweb tutorial page at: http://www.epa.gov/tri/reporting_materials/tutorials/tutorial_index.html

If a facility submitted a hardcopy TRI Form R/A to meet the July 1 deadline because the reporting facility could not process their electronic signature agreements (ESA) on time, should their certifying official still certify electronically after the July 1 deadline?

Yes, their certifying official should certify any pending TRI forms in CDX after the ESA approval has been processed by the Data Processing Center after the July 1 deadline. Paper forms are not encouraged for the submission of TRI data to EPA because of the likelihood of the presence of errors. However, facilities may consider submitting a hard-copy TRI Form R/A if approval of the certifying official's Electronic Signature Agreement (ESA) has not been processed by EPA by the July 1 deadline. EPA discourages facilities from submitting hard-copy forms because of the potential of reporting errors that will force EPA to issue a separate Notice of Significant Error (NOSE) to your facility to correct them. However, this situation is less likely to occur in RY 2012 with the use of the new LexisNexis ESA approval method that will allow for real-time processing of the ESA form. This new ESA approval alternative should discourage reporting facilities from mailing hardcopy TRI forms instead of transmitting, certifying and submitting their electronic submission. EPA strongly encourages your facility's certifying official to log into CDX (using the link stated above) once their Electronic Signature Agreement (ESA) has been approved by EPA and certify any pending submission(s) because your electronic submission(s) should have already passed the validation/data error checks in TRI-MEweb, minimizing any chance of errors in your forms. Your certifying official should electronically certify

any pending TRI submission even after the July 1 deadline. *If a facility certifies any pending electronic submission(s) after their paper form(s) has been received and postmarked by the TRI Data Processing Center, the electronic submission(s) will be processed and will retain the postmarked date of the previously filed paper form(s).*

A.6 When the TRI Report Must Be Submitted

As specified in EPCRA Section 313, the report for any calendar year must be submitted on or before midnight on July 1 of the following year whether using Form R or Form A. If the reporting deadline falls on a Saturday or Sunday, EPA will accept the forms which are postmarked on the following Monday (i.e., the next business day). If your reporting facility submitted TRI forms using TRI-MEweb via the Central Data Exchange (CDX), the preparer will receive their electronic Facility Data Profile (eFDP) in an expedited fashion under the "**eFDP**" tab within the TRI-MEweb application. Any voluntary revision to a report can be submitted anytime during the calendar year for the current or any previous reporting year. However, voluntary revisions for the current reporting year should be submitted by July 31 in order to be included in that year's TRI National Analysis. Always remember to review your eFDP report to verify that data was entered correctly and no addition errors were found after processing by EPA's DPC. The eFDP report is a receipt of the information your reporting facility has submitted to EPA. If the Technical Contact provided an email address in the Form R/Form A, they will receive an email notifying them when their eFDP has been updated and published in TRI-MEweb for review. If you have questions regarding your eFDP, please send an e-mail to tri.efdp@epacdx.net or call 703-227-7644.

A.7 How to Obtain the TRI Reporting Forms

The TRI Form R, Form R Schedule 1, Form A Certification Statement, and related guidance documents may be obtained from EPA's TRI website at: http://www.epa.gov/tri.

B. How to Determine if Your Facility Must Submit a Form R or Is Eligible to Use Form A

This section will help you determine whether you must submit an EPCRA Section 313 report (EPA Form R or Form A Certification Statement). This section discusses EPCRA Section 313 reporting requirements such as the number of full-time employees, primary NAICS code, and chemical activity threshold quantities. The EPCRA Section 313 chemicals and chemical categories subject to reporting are listed in Table II (also see 40 CFR 372.65). (See Figure 1 for more information.)

B.1 Full-Time Employee Determination

The number of full-time employees is dependent only upon the total number of hours worked by all employees and other individuals (e.g., contractors) for the facility during the calendar year and not the number of persons working. Therefore, a full-time employee, for purposes of EPCRA Section 313 reporting, is defined as 2,000 work hours per year. When making the full-time employee determination, the facility must consider all paid vacation and sick leave used as hours worked by each employee. In addition, EPA interprets the hours worked by an employee to include paid holidays. To determine the number of full-time employees working for your facility, add up the hours worked by all employees during the calendar year, including contract employees and sales and support staff working for the facility, and divide the total by 2,000 hours. The result is the number of full-time employees. In other words, if the total number of hours worked by all employees for your facility is 20,000 hours or more, your facility meets the ten employee threshold.

Examples:

- A facility consists of 11 employees who each worked 1,500 hours for the facility in a calendar year. Consequently, the total number of hours worked by all employees for the facility during the calendar year is 16,500 hours. The number of full-time employees for this facility is equal to 16,500 hours divided by 2,000 hours per full-time employee, or 8.3 full-time employees. Therefore, even though 11 persons worked for this facility during the calendar year, the number of hours worked is equivalent to 8.3 full-time employees. This facility does not meet the employee criteria and is not subject to EPCRA Section 313 reporting.

- Another facility consists of six workers and three sales staff. The six workers each worked 2,000 hours for the facility during the calendar year. The sales staff also each worked 2,000 hours during the calendar year although they may have been on the road half of the year. In addition, five contract employees were hired for a period during which each worked 400 hours for the facility. The total number of hours is equal to the time worked by the workers (12,000 hours), plus the time worked by the sales staff for the facility (6,000 hours), plus the time worked by the contract employees (2,000 hours), or 20,000 hours. Dividing the 20,000 hours by 2,000 yields 10 full-time employees. This facility has met the full-time employee criteria and may be subject to reporting if the other criteria are met.

B.2 Primary NAICS Code Determination

Beginning with 2006 EPCRA Section 313 reporting, the TRI Program requires North American Industry Classification System (NAICS) codes instead of Standard Industrial Classification (SIC) codes. Please refer to the TRI Program's final rule titled Community Right-to-Know; Toxic Chemical Release Reporting Using North American Industry Classification System (NAICS) published in the *Federal Register* on June 6, 2006 (71 FR 32464).

EPA published a final rule on June 9, 2008 to incorporate 2007 Office of Management and Budget (OMB) revisions and other corrections to the NAICS codes used for TRI Reporting. [Federal Register (FR) notice 73 FR 32466.] With this rule, facilities are required to use 2007 NAICS codes on TRI reporting forms.

The full list of NAICS codes for facilities that must report to TRI (including exceptions and/or limitations) if all other threshold determinations are met can be found in Table I and also at the TRI website at: http://www.epa.gov/tri/lawsandregs/naic/ncodes.htm. The facility should determine its own NAICS code(s), based on its activities on-site using the NAICS Manual and by conducting NAICS keyword and NAICS 2 to 6-digit code searches on the Census Bureau website at: http://www.census.gov/eos/www/naics/. For purposes of EPCRA Section 313 reporting, state assigned codes should not be used if they differ from codes assigned using the NAICS Manual.

The NAICS 2012 Manual is available from the National Technical Information Service (NTIS) website at: http://www.ntis.gov/about/index.aspx.

Paperwork Reduction Act Notice: The annual public burden related to the Form R, which is approved under OMB Control No. 2025-0009, is estimated to average 35.71 hours per response for a facility filing a report on one chemical. The annual public burden related to the Form A, which is also approved under OMB Control No. 2025-0009, is estimated to average 21.96 hours per response for a facility filing a report on one chemical.

Burden means the total time, effort, or financial resources expended by persons to generate, maintain, retain, or disclose or provide information to or for a Federal agency. This includes the time needed to review instructions; develop, acquire, install, and utilize technology and systems for the purposes of collecting, validating, and verifying information, processing and maintaining information, and disclosing and providing information; adjust the existing ways to comply with any previously applicable instructions and requirements; train personnel to be able to respond to a collection of information; search data sources; complete and review the collection of information; and transmit or otherwise disclose the information. An agency may not conduct or sponsor, and a person is not required to respond to, a collection of information unless it displays a currently valid OMB control number. The OMB control numbers for EPA's regulations are listed in 40 CFR Part 9 and 48 CFR Chapter 15.

Send comments on the Agency's need for this information, the accuracy of the provided burden estimates, and any suggested methods for minimizing respondent burden, including through the use of automated collection techniques, to the Director, Collection Strategies Division, U.S. Environmental Protection Agency (2822), 1200 Pennsylvania Ave., NW, Washington, D.C. 20460; and to the Office of Information and Regulatory Affairs, Office of Management and Budget, 725 17th Street, NW, Washington, DC 20503, Attention: Desk Officer for EPA. Include the EPA ICR number and OMB control number in any correspondence.

The completed forms should be submitted in accordance with the instructions accompanying the form, or as specified in the corresponding regulation.

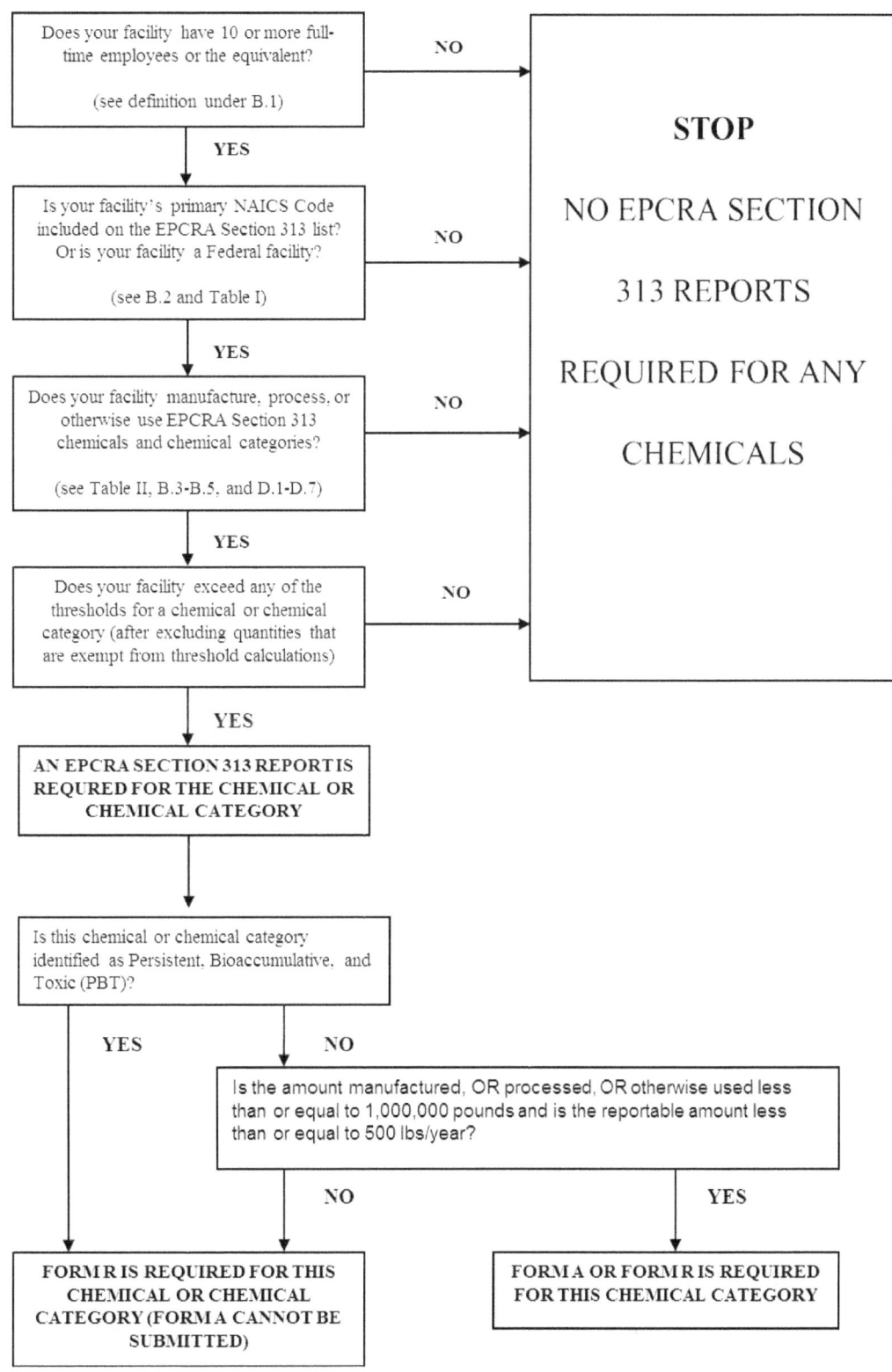

Figure 2. EPCRA Section 313 Reporting Decision Diagram

B.2.a. Auxiliary Facilities

Under the Standard Industrial Classification (SIC) system, an auxiliary facility was defined as one that supported another covered establishment's activities (e.g., research and development laboratories, warehouses, and storage facilities). An auxiliary facility could assume the SIC code of another covered establishment if its primary function was to service that other covered establishment's operations. The North American Industry Classification System (NAICS), that replaces the SIC system for TRI reporting, does not recognize the concept of auxiliary facilities and assigns NAICS codes to all establishments based on economic activity. In its rulemaking, "Toxic Chemical Release Reporting Using North American Industry Classification System," the TRI Program has adopted NAICS for TRI reporting and also the NAICS treatment of former "auxiliary facilities" as entities with their own distinct NAICS code.

B.2.b. Multi-establishment Facilities

Your facility may include multiple establishments that have different NAICS codes. A multi-establishment facility is a facility that consists of two or more distinct and separate economic units. If your facility is a multi-establishment facility, calculate the value added of the products produced, shipped, or services provided from each establishment within the facility and then use the following rule to determine if your facility meets the NAICS code criterion:

- If the total value added of the products produced, shipped, or services provided at establishments with covered NAICS codes is greater than 50 percent of the value added of the entire facility's products and services, the entire facility meets the NAICS code criterion.

- If anyone establishment with a covered NAICS code has a value added of services or products shipped or produced that is greater than any other establishment within the facility (40 CFR Section 372.22(b)(3)) the facility also meets the NAICS code criterion.

The value added of production or service attributable to a particular establishment may be isolated by subtracting the product value obtained from other establishments within the same facility from the total product or service value of the facility. This procedure eliminates the potential for "double counting" production and services in situations where establishments are engaged in sequential production or service activities at a single facility.

Examples include:

- A facility in coating, engraving and allied services has two establishments. The first establishment, a general automotive repair service, is in NAICS code 811113 (SIC 7537), which is not a covered NAICS code. However, the second establishment, a metal paint shop is in NAICS code 332812 (SIC 3479, which is a covered NAICS code). The metal paint shop paints the parts received from general automotive repair service. The facility determines the product is worth $500/unit as received from the general automotive repair service (in non-covered NAICS code 811113) and the value of the product is $1500/unit after processing by the metal paint shop (in covered NAICS code 332812). The value added by the metal paint shop is obtained by subtracting the value of the products from the general automotive repair service from that of the value of the products of the metal paint shop. (In this example, the value added = $1,500/unit - $500/unit = $1,000/unit.) The value added ($1,000/unit) by the establishment in NAICS code 332812 is more than 50 percent of the product value. Therefore, the facility's primary NAICS code is 332812, which is a covered NAICS code.

- A food processing establishment in a facility processes crops grown at the facility in a separate establishment. To determine the value added of the products of each establishment the facility could first determine the value of the crops grown at the agricultural establishment, and then calculate the contribution of the food processing establishment by subtracting the crop value from the total value of the product shipped from the processing establishment (value of product shipped from processing - crop value = value of processing establishment).

A covered multi-establishment facility must make EPCRA Section 313 chemical threshold determinations and, if required, report all relevant information about releases and other waste management activities, and source reduction activities associated with an EPCRA Section 313 chemical **for the entire facility**, even from establishments that are not in covered NAICS codes. EPA realizes, however, that certain establishments in a multi-establishment facility can be, for all practical purposes, separate and distinct business units. Therefore, while threshold determinations must be made for the entire facility, individual establishments which compose the entire facility may report their individual releases and other waste management activities separately. However, the total releases and other waste management quantities for the entire facility must be represented by the sum of the releases and other quantities managed as waste reported by each of the separate establishments.

B.2.c. Property Owners

You are not required to report if you merely own real estate on which a facility covered by this rule is located; that is, you have no other business interest in the operation of that facility (e.g., your company owns an industrial park). The operator of that facility, however, is subject to reporting requirements.

B.3 Activity Determination

B.3.a. Definitions of Manufacture, Process, and Otherwise Use

Manufacture: The term *"manufacture"* means to produce, prepare, compound, or import an EPCRA Section 313 chemical. (See Part II, Section 3.1 of these instructions for further clarification.)

Import is defined as causing the EPCRA Section 313 chemical to be imported into the customs territory of the United States. If you order an EPCRA Section 313 chemical (or a mixture containing the chemical) from a foreign supplier, then you have imported the chemical when that shipment arrives at your facility directly from a source outside of the United States. By ordering the chemical, you have caused it to be imported, even though you may have used an import brokerage firm as an agent to obtain the EPCRA Section 313 chemical.

Do Not Overlook Coincidental Manufacture

The term *"manufacture"* also includes coincidental production of an EPCRA Section 313 chemical (e.g., as a byproduct or impurity) as a result of the manufacture, processing, otherwise use or disposal of another chemical or mixture of chemicals. In the case of coincidental production of an impurity (i.e., an EPCRA Section 313 chemical that remains in the product that is distributed in commerce), the *de minimis* exemption, discussed in Section B.3.c of these instructions, applies. The *de minimis* exemption does not apply to byproducts (e.g., an EPCRA Section 313 chemical that is separated from a process stream and further processed or disposed of). Certain EPCRA Section 313 chemicals may be manufactured as a result of wastewater treatment or other treatment processes. For example, neutralization of wastewater containing nitric acid can result in the coincidental manufacture of a nitrate compound (solution), reportable as a member of the nitrate compounds category.

Process: The term *"process"* means the preparation of a listed EPCRA Section 313 chemical, after its manufacture, for distribution in commerce. Processing is usually the incorporation of an EPCRA Section 313 chemical into a product (see Part II, Section 3.2 of these instructions for further clarification). However, a facility may process an impurity that already exists in a raw material by distributing that impurity in commerce. Processing includes preparation of the EPCRA Section 313 chemicals in the same physical state or chemical form as that received by your facility, or preparation that produces a change in physical state or chemical form. The term also applies to the processing of a mixture or other trade name product (see Section B.4.b of these instructions) that contains a listed EPCRA Section 313 chemical as one component.

Otherwise Use: The term *"otherwise use"* means any use of an EPCRA Section 313 chemical, including an EPCRA Section 313 chemical contained in a mixture or other trade name product or waste, that is not covered by the terms manufacture or process. Otherwise use of an EPCRA Section 313 chemical does not include disposal, stabilization (without subsequent distribution in commerce), or treatment for destruction unless:

(1) The EPCRA Section 313 chemical that was disposed of, stabilized, or treated for destruction

was received from off-site for the purposes of further waste management;

Or

(2) The EPCRA Section 313 chemical that was disposed of, stabilized, or treated for destruction was manufactured as a result of waste management activities on materials received from off-site for the purposes of waste management activities. Relabeling or redistributing of the EPCRA Section 313 chemical where no repackaging of the EPCRA Section 313 chemical occurs does not constitute an otherwise use or processing of the EPCRA Section 313 chemical. (See 62 FR 23846 and Part II, Section 3.3 of these instructions for further clarification).

Example 1: Coincidental Manufacture

❑ Your company, a nitric acid manufacturer, uses aqueous ammonia in a waste treatment system to neutralize an acidic wastewater stream containing nitric acid. The reaction of ammonia and nitric acid produces a solution of ammonium nitrate. Ammonium nitrate (solution) is reportable under the nitrate compounds category and is manufactured as a byproduct. If the ammonium nitrate is produced in a quantity that exceeds the 25,000-pound manufacturing threshold, the facility must report under the nitrate compounds category.

The aqueous ammonia is considered to be otherwise used and 10 percent of the total aqueous ammonia would be counted towards the 10,000-pound otherwise use threshold. Reports for releases of ammonia must also include 10 percent of the total aqueous ammonia from the solution of ammonium nitrate (see the qualifier for the ammonia listing).

❑ As another example, combustion of coal or other fuel in boilers/furnaces can result in the coincidental manufacture of metal category compounds and sulfuric acid (acid aerosols), hydrochloric acid (acid aerosols), and hydrogen fluoride.

Example 2:   Typical Process and Manufacture Activities

❑ Your company receives toluene, an EPCRA Section 313 chemical, from another facility, and reacts the toluene with air to form benzoic acid, which the company distributes in commerce. Your company processes toluene and manufactures and processes benzoic acid. Benzoic acid, however, is not an EPCRA Section 313 chemical and thus does not trigger reporting requirements.

❑ Your facility combines toluene purchased from a supplier with various materials to form paint which it then sells. Your facility processes toluene.

❑ Your company receives a nickel compound (nickel compounds is a listed EPCRA Section 313 chemical category) as a bulk solid and performs various size-reduction operations (e.g., grinding) before packaging the compound in 50-pound bags, which the company sells. Your company processes the nickel compound.

❑ Your company receives a prepared mixture of resin and chopped fiber to be used in the injection molding of plastic products. The resin contains a listed EPCRA Section 313 chemical that becomes incorporated into the plastic, which the company distributes in commerce. Your facility processes the EPCRA Section 313 chemical.

❑ In the combustion of coal or oil, metal category compounds may be produced from either the parent metal or a metal compound contained in the coal or oil. If a metal undergoes a change of valence, a metal compound is considered to be manufactured. For example, during the combustion process copper in valence state zero changes to copper in valence state +2 in a compound such as copper (II) oxide (CuO). Furthermore, a metallic compound could be transformed to another metallic compound without a change in valency (e.g., copper (II) chloride (CuCl2) is transformed to copper (II) oxide (CuO)). The transformation to a new compound by combustion without a change in valence state is also considered to be "manufactured" for purposes of EPCRA Section 313.

Example 3:   Typical Otherwise Use Activities

❑ When your facility cleans equipment with toluene, you are otherwise using toluene. Your facility also separates two components of a mixture by dissolving one component in toluene, and subsequently recovers the toluene from the process for reuse or disposal. Your facility otherwise uses toluene.

❑ A covered facility receives a waste containing 12,000 pounds of Chemical A, a non-PBT EPCRA Section 313 chemical, from off-site. The facility treats the waste, destroying Chemical A and in the treatment process manufactures 10,500 pounds of Chemical B, another non-PBT EPCRA Section 313 chemical. Chemical B is disposed of on-site. Since the waste containing Chemical A was received from off-site for the purpose of waste management, the amount of Chemical A must be included in the otherwise use threshold determination for Chemical A. The otherwise use threshold for a non-PBT chemical is 10,000 pounds and since the amount of Chemical A exceeds this threshold, all releases and other waste management activities for Chemical A must be reported. Chemical B was manufactured in the treatment of a waste received from off-site. The facility disposed of Chemical B on-site. Since Chemical B was generated from waste received from off-site for treatment for destruction, disposal, or stabilization, the disposal of Chemical B is considered to be an otherwise use. Thus, the amount of Chemical B must be considered in the otherwise use threshold determination. Thus, the reporting threshold for Chemical B has also been exceeded and all releases and other waste management activities for Chemical B must be reported.

B.3.b. Persistent Bioaccumulative Toxic (PBT) Chemicals and Chemical Categories Overview

On October 29, 1999, EPA published a final rule (64 FR 58666) adding certain chemicals and chemical categories to the EPCRA Section 313 list of toxic chemicals and lowering the reporting threshold for persistent bioaccumulative toxic (PBT) chemicals. In addition, on January 17, 2001 EPA published a final rule (66 FR 4500) that classified lead and lead compounds as PBT chemicals and lowered their reporting thresholds. The lower reporting thresholds for lead applies to all lead except when lead is contained in a stainless steel, brass or bronze alloy.

Dioxin and dioxin-like compounds, lead compounds, mercury compounds and polycyclic aromatic compounds (PACs) are the four PBT chemical categories with lower reporting thresholds. The 17 members of the dioxin and dioxin-like compounds category and the 21 members of the PACs category are listed in Table IIc of these instructions. The dioxin and dioxin-like compounds category has the qualifier, "Manufacturing; and the processing or otherwise use of dioxin and dioxin-like compounds if the dioxin and dioxin-like compounds are present as contaminants in a chemical and if they were created during the manufacturing of that chemical."

EPA has added six individual chemicals to the EPCRA Section 313 list of toxic chemicals that also had their thresholds lowered:

- benzo(g,h,i)perylene,
- benzo(j,k)fluorene (fluoranthene),
- 3-methylcholanthrene,
- octachlorostyrene,
- pentachlorobenzene, and
- tetrabromobisphenol A (TBBPA).

Benzo(j,k)fluorene and 3-methyl-cholanthrene were added as members of the polycyclic aromatic compounds (PACs) chemical category.

EPA lowered the reporting thresholds for PBT chemicals to either 100 pounds, 10 pounds, or in the case of the dioxin and dioxin-like compounds chemical category, to 0.1 grams. The table at the beginning of Section B.4 of these instructions lists the applicable manufacture, process, and otherwise use thresholds for the listed PBT chemicals.

EPA eliminated the *de minimis* exemption for all PBT chemicals (except lead when contained in stainless steel, brass or bronze alloy). However, this action does not affect the applicability of the *de minimis* exemption to the supplier notification requirements (40 CFR Section 372.45(d) (1)). In addition, PBT chemicals are ineligible for range reporting for on-site releases and transfers off-site for further waste management. This will not affect the applicability of range reporting of the maximum amount on-site as required by EPCRA Section 313(g).

All releases and other waste management quantities greater than 0.1 pounds of a PBT chemical (except the dioxin and dioxin like compounds chemical category) should be reported at a level of precision supported by the accuracy of the underlying data and estimation techniques on which the estimate is based. If a facility's release or other waste management estimates support reporting an amount that is more precise than whole numbers, then the more precise amount should be reported.

For the dioxin and dioxin-like compounds chemical category, which has a reporting threshold of 0.1 grams, facilities need only report all release and other waste management quantities greater than 100 micrograms (i.e., 0.0001 grams). Notwithstanding the numeric precision used when determining reporting eligibility thresholds, facilities should report on the Form R to the level of accuracy that their data supports, up to seven digits to the right of the decimal. EPA's reporting software and data management systems support data precision to seven digits to the right of the decimal. If a facility has information on the individual members of the dioxin and dioxin-like compounds category they will also need to file a Form R Schedule 1 (see instructions in Section D).

Lead and Lead Compounds

Lead and lead compounds are classified as PBT chemicals and are subject to the lower manufacturing, processing and otherwise use threshold of 100 pounds. However, when lead is contained in stainless steel, brass, or bronze alloys it remains subject to the higher 25,000 pound manufacturing and processing thresholds and the 10,000 pound otherwise use threshold.

Listed below are some important guidelines to use when calculating threshold and release and other waste management quantities for lead and lead compounds:

1) quantities of lead not contained in stainless steel, brass or bronze alloy are applied to both the 100 pound threshold and the 25,000/10,000 pound thresholds;

2) quantities of lead that are contained in stainless steel, brass or bronze alloys are only applied toward the 25,000/10,000 pound thresholds;

3) a facility may take the *de minimis* exemption for those quantities of lead in stainless steel, brass, or bronze alloys that meet the *de minimis* standard (e.g., manufactured as an impurity). Accordingly, the *de minimis* exemption may be considered for quantities of lead in stainless steel, brass, or bronze alloys but it may not be considered for lead not in stainless steel, brass, or bronze alloys;

4) If a facility exceeds the 100-pound threshold for lead other than in stainless steel, brass, or bronze alloys, the facility may not apply Form A eligibility for non-PBTs, range reporting in Sections 5 and 6 of the Form R or the use of whole numbers and 2 significant digits to any of the lead they report. If a facility that exceeds the 25,000/10,000 pound threshold for lead in stainless steel, brass, or bronze alloy without tripping the 100-pound threshold for non-alloyed lead, the facility may consider the Form A requirements for non-PBTs, range reporting in Sections 5 and 6 of the Form R, and the use of whole numbers and 2 significant digits.

B.3.c. Activity Exemptions

Otherwise Use Exemptions. Certain otherwise uses of listed EPCRA Section 313 chemicals are specifically exempted:

- Otherwise use as a structural component of the facility;

- Otherwise use in routine janitorial or facility grounds maintenance;

- Personal uses by employees or other persons;

- Otherwise use of products containing EPCRA Section 313 chemicals for the purpose of maintaining motor vehicles operated by the facility; and

- Otherwise use of EPCRA Section 313 chemicals contained in intake water (used for processing or non-contact cooling) or in intake air (used either as compressed air or for combustion).

The exemption of an EPCRA Section 313 chemical otherwise used 1) as a structural component of the facility; or 2) in routine janitorial or facility grounds maintenance; or 3) for personal use by an employee cannot be taken for activities involving process related equipment.

Articles Exemption. EPCRA Section 313 chemicals contained in articles that are processed or otherwise used at a covered facility are exempt from threshold determinations and release and other waste management calculations. The exemption applies when the facility receives the article from another facility or when the facility produces the article itself. The exemption applies only to the quantity of EPCRA Section 313 chemical present in the article. If the EPCRA Section 313 chemical is manufactured (including imported), processed, or otherwise used at the covered facility other than as part of the article, in excess of an applicable threshold quantity, the facility is required to report that use of a chemical (40 CFR Section 372.38(b)). For an EPCRA Section 313 chemical in an item to be exempt as part of the article, the item must meet all the following criteria in the EPCRA Section 313 article definition; that is, it must be a manufactured item (1) which is formed to a specific shape or design during manufacture, (2) which has end use functions dependent in whole or in part upon its shape or design during end use, and (3) which does not release a toxic chemical under normal conditions of processing or use of the item at the facility.

If the processing or otherwise use of all like items results in a total release of 0.5 pound or less of an EPCRA Section 313 chemical in a reporting year to any environmental medium, EPA will allow this release to be rounded to zero, and the manufactured items retain their article status. The 0.5 pound threshold does not apply to each individual article, but applies to the sum of all releases from processing or otherwise use of all like articles. If all the releases of like articles over a reporting year are completely captured and recycled/reused on-site or off-site, those items retain their article status. Any amount that is released and is not recycled/reused will count toward the 0.5 pound per year cut off value.

The articles exemption applies to the normal processing or use of articles. This exemption does not apply to the manufacture of the article. EPCRA Section 313 chemicals incorporated into articles produced at a facility must be factored into threshold determinations and release and other waste management calculations.

If, in the course of processing or use, an item retains its initial thickness or diameter, in whole or in part, it meets the first part (i.e., it must be a manufactured item which is formed to a specific shape or design during manufacture) of the article definition. If the item's basic dimensional characteristics are totally altered during processing or otherwise use, the item does not meet the first part of the definition. An example of items that do not meet the definition would be items that are cold extruded, such as lead ingots, which are formed into

wire or rods. On the other hand, cutting a manufactured item into pieces that are recognizable as the article would not change the original dimensions as long as the diameter or the thickness of the item remained the same; the articles exemption would continue to apply. Metal wire may be bent and sheet metal may be cut, punched, stamped, or pressed without losing their article status as long as the diameter of the wire or tubing or the thickness of the sheet is not totally changed.

What constitutes a release of an EPCRA Section 313 chemical is important since processing or otherwise use of articles that result in a release to the environment (or more than 0.5 pounds) negate the article status and precludes eligibility for the exemption. Cutting, grinding, melting, or other processing of manufactured items could result in a release of an EPCRA Section 313 chemical during normal conditions of processing or otherwise use and therefore negate the exemption as articles.

Example 4: Articles Exemption

❑ Nickel that is incorporated into a brass doorknob is processed to manufacture the brass doorknob, and therefore must be counted toward threshold determinations and release and other waste management calculations. However, the use of the brass doorknobs elsewhere in the facility does not have to be counted. Disposal of the brass doorknob after its use does not constitute a "release;" thus, the brass doorknob remains an article.

❑ If an item used in the facility is fragmented, the item is still an article if those fragments being discarded remain identifiable as the article (e.g., recognizable pieces of a cylinder, pieces of wire). For instance, an eight-foot piece of wire is cut into two four-foot pieces of wire, without releasing any EPCRA Section 313 chemicals. Each four-foot piece is identifiable as a piece of wire; therefore, the article status for these pieces of wire remains intact.

❑ EPCRA Section 313 chemicals received in the form of pellets are not articles because the pellet form is simply a convenient form for further processing of the material.

De Minimis **Exemption.** The *de minimis* exemption allows facilities to disregard certain minimal concentrations of non-PBT chemicals in mixtures or other trade name products when making threshold determinations and release and other waste management calculations. The *de minimis* exemption does not apply to the manufacture of an EPCRA Section 313 chemical except if that EPCRA Section 313 chemical is manufactured as an impurity and remains in the product distributed in commerce, or if the EPCRA Section 313 chemical is imported below the appropriate *de minimis* level. The *de minimis* exemption does not apply to a byproduct manufactured coincidentally as a result of manufacturing, processing, otherwise use, or any waste management activities. The *de minimis* exemption does not apply to any PBT chemical (except lead when it is contained in stainless steel, brass or bronze alloy) or PBT chemical category. A list of PBT chemicals may be found in Section B.4 of these instructions.

When determining whether the *de minimis* exemption applies to an EPCRA Section 313 chemical, the owner/operator must consider the concentration of the non-PBT EPCRA Section 313 chemical in mixtures and other trade name products. If the non-PBT EPCRA Section 313 chemical in a mixture or other trade name product is manufactured as an impurity, imported, processed, or otherwise used and is below the appropriate *de minimis* concentration level, then the quantity of the non-

PBT EPCRA Section 313 chemical in that mixture or other trade name product does not have to be applied to threshold determinations nor included in release or other waste management determinations. If a non-PBT EPCRA Section 313 chemical in a mixture or other trade name product is below the appropriate *de minimis* level, all releases and other waste management activities associated with the EPCRA Section 313 chemical in that mixture or other trade name product are exempt from EPCRA Section 313 reporting. It is possible to meet an activity (e.g., processing) threshold for an EPCRA Section 313 chemical on a facility wide basis, but not be required to calculate releases or other waste management quantities associated with a particular process because that process involves only mixtures or other trade name products containing the non-PBT EPCRA Section 313 chemical below the *de minimis* level.

EPA interprets the *de minimis* exemption such that once a non-PBT EPCRA Section 313 chemical concentration is at or above the appropriate *de minimis* level in the mixture or other trade name product threshold determinations and release and other waste management calculations must be made, even if that chemical later falls below the *de minimis* level in the same mixture or other trade name product. Thus, EPA considers reportable all releases and other quantities managed as waste that occur after the *de minimis* level has been met or exceeded. If an EPCRA Section 313 chemical in a mixture or other trade name product at or above *de minimis* is brought on-site, the *de minimis* exemption never applies.

De minimis levels for non-PBT EPCRA Section 313 chemicals and chemical categories are set at concentration levels of either 1 percent or 0.1 percent; PBT chemicals and chemical categories do not have *de minimis* levels with regard to this exemption. The 0.1 percent *de minimis* levels are dictated by determinations made by the National Toxicology Program (NTP) in its Annual Report on Carcinogens, the International Agency for Research and Cancer (IARC) in its Monographs, or 29 CFR part 1910, subpart Z. Therefore, once a non-PBT chemical's status under NTP, IARC, or 29 CFR part 1910, subpart Z indicates that the chemical is a carcinogen or potential carcinogen, the reporting facility may disregard levels of the chemical below the 0.1 percent *de minimis* concentration provided that the other criteria for the *de minimis* exemption are met. *De minimis* levels for chemical categories apply to the total concentration of all chemicals in the category within a mixture, not the concentration of each individual category member within the mixture.

De Minimis Application to the Processing or Otherwise Use of a Mixture

The *de minimis* exemption applies to the processing or otherwise use of a non-PBT EPCRA Section 313 chemical in a mixture. Threshold determinations and release and other waste management calculations begin at the point where the chemical meets or exceeds the *de minimis* level. If a non-PBT EPCRA Section 313 chemical is present in a mixture at a concentration below the *de minimis* level, this quantity of the substance does not have to be included for threshold determinations, release and other waste management reporting, or supplier notification requirements. The exemption will apply as long as the mixture containing *de minimis* amounts of a non-PBT EPCRA Section 313 chemical never equals or goes above the *de minimis* limit.

Example 5: *De Minimis* **Applications to Process and Otherwise Use Scenarios for Non-PBT Chemicals**

There are many cases in which the *de minimis* "limit" is crossed or re-crossed by non-PBT chemicals within a process or otherwise use scenario. The following examples are meant to illustrate these complex reporting scenarios.

Increasing Concentration To or Above *De Minimis* **Levels During Processing for Non-PBT Chemicals**

A manufacturing facility receives toluene that contains chlorobenzene at a concentration below its *de minimis* limit. Through distillation, the chlorobenzene content in process streams is increased over the *de minimis* concentration of 1 percent. From the point at which the chlorobenzene concentration equals 1 percent in process streams, the amount present must be factored into threshold determinations and release and other waste management estimates. The facility does not need to consider the amount of chlorobenzene in the raw material when below *de minimis* levels, i.e., prior to distillation to 1 percent, when making threshold determinations. The facility does not have to report emissions of chlorobenzene from storage tanks or any other equipment associated with that specific process where the chlorobenzene content is less than 1 percent.

Fluctuating Concentration During Processing for Non-PBT Chemicals

A manufacturer produces an ink product that contains toluene, an EPCRA Section 313 chemical, below the *de minimis* level. The process used causes the percentage of toluene in the mixture to fluctuate: it rises above the *de minimis* level for a time but drops below the level as the process winds down. The facility must consider the chemical toward threshold determinations from the point at which it first equals the *de minimis* limit. Once the *de minimis* limit has been met the exemption cannot be taken.

Concentration Ranges Straddling the *De Minimis* **Value**

There may be instances in which the concentration of a non-PBT chemical is given as a range straddling the *de minimis* limit. Example 6 illustrates how the *de minimis* exemption should be applied in such a scenario.

***De Minimis* Application in the Manufacture of the Listed Chemical in a Mixture**

The *de minimis* exemption generally does not apply to the manufacturing of an EPCRA Section 313 chemical. However, the *de minimis* exemption may apply to mixtures and other trade name products containing non-PBT EPCRA Section 313 chemicals that are imported into the United States. (See Example 5)

The exemption also applies to non-PBT EPCRA Section 313 chemicals that are manufactured as impurities that remain in the product distributed in commerce below the *de minimis* levels. The amount remaining in the product is exempt from threshold determinations. If the chemical is separated from the final product, it cannot qualify for the exemption. Any amount that is separated, or is separate, from the product, is considered a byproduct and is subject to threshold determinations and release and other waste management calculations. Any amount of an EPCRA Section 313 chemical that is manufactured in a waste stream must be considered toward threshold determinations and release and other waste management calculations and accounted for on Form R even if that chemical is manufactured below the *de minimis* level.

The *de minimis* exemption also does not apply to situations where a toxic chemical in waste is diluted to below the *de minimis* level.

Example 6: Concentration Ranges Straddling the *De Minimis* Value

Scenario 1: A facility processes 8,000,000 pounds of a mixture containing 0.25 to 1.25 percent manganese. Manganese is eligible for the *de minimis* exemption at concentrations up to 1 percent. The amount of mixture subject to reporting is the quantity containing manganese at or above the *de minimis* concentration:

$$[(8,000,000) \times (1.25\% - 0.99\%)] \div (1.25\% - 0.25\%)$$

The average concentration of manganese that is not exempt (above the *de minimis*) is:

$$(1.25\% + 1.00\%) \div (2)$$

Therefore, the amount of manganese that is subject to threshold determination and release and other waste management estimates is:

$$\left[\frac{(8,000,000) \times (1.25\% - 0.99\%)}{(1.25\% - 0.25\%)} \right] \times \left[\frac{(1.25\% + 1.00\%)}{(2)} \right] = 23,400 \, pounds$$

= 23,400 pounds manganese (which is below the processing threshold for manganese)

In this scenario, because the facility's information pertaining to manganese was available to two decimal places, 0.99 was used to determine the amount below the *de minimis* concentrations. If the information was available to one decimal place, 0.9 should be used, as in the scenario below.

Scenario 2: As in the previous example, manganese is present in a mixture, of which 8,000,000 pounds is processed. The MSDS states the mixture contains 0.2 percent to 1.2 percent manganese. The amount of mixture subject to reporting (at or above *de minimis* limit) is:

$$[(8,000,000) \times (1.2\% - 0.9\%)] \div (1.2\% - 0.2\%)$$

The average concentration of manganese that is not exempt (at or above *de minimis* limit) is:

$$(1.2\% + 1.0\%) \div (2)$$

Therefore, the amount of manganese that is subject to threshold determinations and release and other waste management estimates is:

$$\left[\frac{(8,000,000) \times (1.2\% - 0.9\%)}{(1.2\% - 0.2\%)} \right] \times \left[\frac{(1.2\% + 1.0\%)}{(2)} \right] = 26,400 \, pounds$$

= 26,400 pounds manganese (which is above the processing threshold for manganese)

Example 7: *De Minimis* **Application in the Manufacture of a Toxic Chemical in a Mixture**

Manufacture as a Product Impurity

Toluene 2,4 diisocyanate reacts with trace amounts of water to form trace quantities of 2,4-diaminotoluene. The resulting product contains 99 percent toluene 2,4-diisocyanate and 0.05 percent 2,4-diaminotoluene. The 2,4 diaminotoluene would not be subject to EPCRA Section 313 reporting nor would supplier notification be required because the concentration of 2,4- diaminotoluene is below its *de minimis* limit of 0.1 percent in the product.

Manufacture as a Commercial Byproduct and Impurity

Chloroform is a reaction byproduct in the production of carbon tetrachloride. It is removed by distillation to a concentration of less than 150 ppm (0.0150 percent) remaining in the carbon tetrachloride. The separated chloroform at 90 percent concentration is sold as a byproduct. Chloroform is subject to a 0.1 percent (1000 ppm) *de minimis* limit. Any amount of chloroform manufactured and separated as byproduct must be included in threshold determinations because EPA does not interpret the *de minimis* exemption to apply to the manufacture of a chemical as a byproduct. Releases of chloroform prior to and during purification of the carbon tetrachloride must be reported. The *de minimis* exemption can, however, be applied to the chloroform remaining in the carbon tetrachloride as an impurity. Because the concentration of chloroform remaining in the carbon tetrachloride is below the *de minimis* limit, this quantity of chloroform is exempt from threshold determinations, release and other waste management reporting, and supplier notification.

Manufacture as a Waste Byproduct

A small amount of formaldehyde is manufactured as a reaction byproduct during the production of phthalic anhydride. The formaldehyde is separated from the phthalic anhydride as a waste gas and burned, leaving no formaldehyde in the phthalic anhydride. The amount of formaldehyde produced and removed must be included in threshold determinations and release and other waste management estimates even if the formaldehyde were present below the *de minimis* level in the process stream where it was manufactured or in the waste stream to which it was separated because EPA does not interpret mixtures and trade name products to includes wastes.

Laboratory Activities Exemption. EPCRA Section 313 chemicals that are manufactured, processed, or otherwise used in a laboratory at a covered facility under the direct supervision of a technically qualified individual do not have to be considered for threshold determinations and release and other waste management calculations. However, pilot plant scale and specialty chemical production does not qualify for this laboratory activities exemption, nor does the use of EPCRA Section 313 chemicals for laboratory support activities, such as the use of chemicals for equipment maintenance.

Coal Extraction Activities Exemption. If an EPCRA Section 313 chemical is manufactured, processed, or otherwise used in extraction by facilities in NAICS codes 212111, 212112 and 212113, a person is not required to consider the quantity of the EPCRA Section 313 chemical so

manufactured, processed, or otherwise used when considering threshold determinations and release and other waste management calculations (see Example 8). Reclamation activities occurring simultaneously with coal extraction activities (e.g., cast blasting) are included in the exemption. However, otherwise use of ash, waste rock, or fertilizer for reclamation purposes are not considered part of extraction; non-exempt amounts of EPCRA Section 313 chemicals contained in these materials must be considered toward threshold determinations and release and other waste management calculations.

Metal Mining Overburden Exemption. If an EPCRA Section 313 chemical that is a constituent of overburden is processed or otherwise used by facilities in NAICS codes 212221, 212222, 212231, 212234, and 212299, a person is not required to

---

consider the quantity of the EPCRA Section 313 chemical so processed or otherwise used when considering threshold determinations and release and other waste management calculations.

For purposes of EPCRA Section 313 reporting, overburden is the unconsolidated material that overlies a deposit of useful material or ore. It does not include any portion of the ore or waste rock.

Example 8: Coal mining extraction activities

Included among these are explosives for blasting operations, solvents, lubricants, and fuels for extraction related equipment maintenance and use, as well as overburden and mineral deposits. The EPCRA Section 313 chemicals contained in these materials are exempt from threshold determinations and release and other waste management calculations, when manufactured, processed or otherwise used during extraction activities at coal mines.

B.4 Threshold Determinations

EPCRA Section 313 reporting is required if threshold quantities are exceeded. Separate thresholds apply to the amount of the EPCRA Section 313 chemical that is manufactured, processed or otherwise used.

You must submit a report for any EPCRA Section 313 chemical that is not listed as a PBT chemical and which is manufactured or processed at your facility in excess of the following threshold:

- 25,000 pounds per toxic chemical or category over the calendar year.

- You must submit a report for any EPCRA Section 313 chemical which is not listed as a PBT chemical and that is otherwise used at your facility in excess of 10,000 pounds per toxic chemical or category over the calendar year.

You must submit a report for any EPCRA Section 313 chemical that is listed as a PBT chemical and which is manufactured, processed or otherwise used at your facility above the designated threshold for that chemical.

The PBT chemical names, CAS numbers and their reporting thresholds are listed in the table below. See Table IIc of these instructions for lists of individual members of the dioxin and dioxin-like compounds chemical category and the polycyclic aromatic compounds (PACs) chemical category.

Chemical or chemical category name	CAS number or chemical category code	Threshold (pounds, unless noted otherwise)
Aldrin	309-00-2	100
Benzo[g,h,i]perylene	191-24-2	10
Chlordane	57-74-9	10
Dioxin and dioxin-like compounds category (manufacturing; and the processing or otherwise use of dioxin and dioxin-like compounds category if the dioxin and dioxin-like compounds are present as contaminants in a chemical and if they were created during the manufacturing of that chemical)	N150	0.1 gram
Heptachlor	76-44-8	10
Hexachlorobenzene	118-74-1	10
Isodrin	465-73-6	10
Lead (this lower threshold does not apply to lead when it is contained in stainless steel, brass or bronze alloy)	7439-92-1	100
Lead compounds	N420	100
Mercury	7439-97-6	10
Mercury compounds	N458	10
Methoxychlor	72-43-5	100
Octachlorostyrene	29082-74-4	10
Pendimethalin	40487-42-1	100
Pentachlorobenzene	608-93-5	10
Polychlorinated biphenyls (PCBs)	1336-36-3	10
Polycyclic aromatic compounds category (PACs)	N590	100
Tetrabromobisphenol A	79-94-7	100
Toxaphene	8001-35-2	10
Trifluralin	1582-09-8	100

B.4.a. How to Determine if Your Facility Has Exceeded Thresholds

To determine whether your facility has exceeded an EPCRA Section 313 reporting threshold, compare quantities of EPCRA Section 313 chemicals that you manufacture, process, or otherwise use to the respective thresholds for those activities. A worksheet is provided in Figure 3A to assist facilities in determining whether they exceed any of the reporting thresholds for non-PBT chemicals; Figures 3B-D provide worksheets for PBT chemicals. (The worksheets can be found at the end of section B.5.) These worksheets also provide a format for maintaining reporting facility records. Use of these worksheets is not required and the completed worksheet(s) should not accompany Form R reports submitted to EPA and the state or tribe.

Complete the appropriate worksheet for each EPCRA Section 313 chemical or chemical category. Base your threshold determination for EPCRA Section 313 chemicals with qualifiers only on the quantity of the EPCRA Section 313 chemical satisfying the qualifier.

Use of the worksheets is divided into three steps:

- *Step 1* allows you to record the gross amount of the EPCRA Section 313 chemical or chemical category involved in activities throughout the facility. Pure forms as well as the amounts of the EPCRA Section 313 chemical or chemical category present in mixtures or other trade name products must be considered. The types of activity (i.e., manufacturing, processing, or otherwise using) for which the EPCRA Section 313 chemical is used must be identified because separate thresholds apply to each of these activities. A record of the information source(s) used should be kept. Possible information sources include purchase records, inventory data, and calculations by a process engineer. The data collected in Step 1 will be totaled for each activity to identify the overall amount of the EPCRA Section 313 chemical or chemical category manufactured (including imported), processed, or otherwise used.

- *Step 2* allows you to identify uses of the EPCRA Section 313 chemical or chemical category that were included in Step 1 but are exempt under

EPCRA Section 313. Do not include in Step 2 exempt quantities of the EPCRA Section 313 chemical not included in the calculations in Step 1. For example, if Freon contained in the building's air conditioners was not reported in Step 1, you would not include the amount as exempt in Step 2. Step 2 is intended for use when a quantity or use of the EPCRA Section 313 chemical is exempt while other quantities require reporting. Note the type of exemption for future reference. Also identify, if applicable, the fraction or percentage of the EPCRA Section 313 chemical present that is exempt. Add the amounts in each activity to obtain a subtotal for exempted amounts of the EPCRA Section 313 chemical or chemical categories at the facility.

- *Step 3* involves subtracting the result of Step 2 from the results of Step 1 for each activity. Compare this net sum to the applicable activity threshold. If the threshold is exceeded for any of the three activities, a facility must submit a Form R for that EPCRA Section 313 chemical or chemical category. Do not sum quantities of the EPCRA Section 313 chemical that are manufactured, processed, and otherwise used at your facility, because each of these activities requires a separate threshold determination. For example, if in a calendar year you processed 20,000 pounds of a non-PBT EPCRA Section 313 chemical and you otherwise used 6,000 pounds of that same chemical, your facility has not exceeded any applicable threshold and thus is not required to report for that chemical.

Worksheets should be retained to document your determination for reporting or not reporting, but should not be submitted with the report.

You must submit a report if you exceed any threshold for any EPCRA Section 313 chemical or chemical category. For example, if your facility processes 22,000 pounds of a non-PBT EPCRA Section 313 chemical and also otherwise uses 16,000 pounds of that same chemical, it has exceeded the otherwise use threshold (10,000 pounds for a non-PBT chemical) and your facility must report even though it did not exceed the process threshold (25,000 pounds for a non-PBT chemical). In preparing your reports, you must consider all non-exempted activities and all releases and other waste management quantities of the EPCRA Section 313 chemical from your facility,

not just releases and other waste management quantities from the otherwise use activity.

Also note that threshold determinations are based upon the actual amounts of an EPCRA Section 313 chemical manufactured, processed, or otherwise used over the course of the calendar year. The threshold determination may not relate to the amount of an EPCRA Section 313 chemical brought on-site during the calendar year. For example, if a stockpile of 100,000 pounds of a non-PBT EPCRA Section 313 chemical is present on-site but only 20,000 pounds of that chemical is applied to a process, only the 20,000 pounds processed is counted toward a threshold determination, not the entire 100,000 pounds of the stockpile.

B.4.b.   Threshold Determinations for On-Site Reuse Operations

Threshold determinations of EPCRA Section 313 chemicals that are reused at the facility are based only on the amount of the EPCRA Section 313 chemical that is added during the year, not the total volume in the system. For example, a facility operates a refrigeration unit that contains 15,000 pounds of anhydrous ammonia at the beginning of the year. The system is charged with 2,000 pounds of anhydrous ammonia during the year. The facility has therefore "otherwise used" only 2,000 pounds of anhydrous ammonia, a non-PBT EPCRA Section 313 chemical, which is below the otherwise use threshold for anhydrous ammonia and is not required to report (unless there are other "otherwise use" activities of ammonia, that when taken together, exceed the reporting threshold). If, however, the whole refrigeration unit was recharged with 15,000 pounds of anhydrous ammonia during the year, then the facility would have exceeded the otherwise use threshold, and would be required to report.

This does not apply to EPCRA Section 313 chemicals "recycled" or "reused" off-site and returned to a facility. Such EPCRA Section 313 chemicals returned to a facility are treated as the equivalent of newly purchased material for purposes of EPCRA Section 313 threshold determinations.

B.4.c.   Threshold Determinations for Ammonia

The listing for ammonia includes the modifier "includes   anhydrous   ammonia   and   aqueous ammonia from water dissociable ammonium salts and other sources; 10 percent of total aqueous ammonia is reportable under this listing." The qualifier for ammonia means that anhydrous forms of ammonia are 100 percent reportable and aqueous forms are limited to 10 percent of total aqueous ammonia. Therefore, when determining threshold quantities, 100 percent of anhydrous ammonia is included but only 10 percent of total aqueous ammonia is included. If any ammonia evaporates from aqueous ammonia solutions, 100 percent of the evaporated ammonia is included in threshold determinations.

For example, if a facility processes aqueous ammonia, it has processed 100 percent of the aqueous ammonia in that solution. If the ammonia remains in solution, then 10 percent of the total aqueous ammonia is counted towards the threshold. If there are any evaporative losses of anhydrous ammonia, then 100 percent of those losses must be counted towards the processing threshold. If the manufacturing, processing, or otherwise use threshold for the ammonia listing is exceeded, the facility must report 100 percent of these evaporative losses in Sections 5 and 8 of the Form R.

B.4.d. Threshold Determinations for Chemical Categories

A number of chemical compound categories are subject to reporting. See Table IIc for a listing of these EPCRA Section 313 chemical categories. When preparing threshold determinations for one of these EPCRA Section 313 chemical categories, all individual members of a category that are manufactured, processed, or otherwise used must be counted. Where generic names are used at a facility, threshold determinations should be based on CAS numbers. For example, Poly-Solv EB does not appear among the reportable chemicals in Table IIa or IIb but its CAS number indicates Poly-Solv EB is a synonym for ethylene glycol mono-n-butyl ether, a member of the certain glycol ethers chemical category (code N230). For chemical compound categories, threshold determinations must be made separately for each of the three activities. Do not include in these threshold determinations for a category any chemicals that are also individually listed EPCRA Section 313 chemicals (see Table IIa or IIb). Individually listed EPCRA Section 313 chemicals are subject to their own individual threshold determination.

Organic Compounds

For the organic compound categories, you are required to account for the entire weight of all compounds within a specific compound category (e.g., glycol ethers) at the facility for BOTH the threshold determination and release and other waste management estimates.

Metal Category Compounds

Threshold determinations for metal category compounds present a special case. If, for example, your facility processes several different nickel compounds, base your threshold determination on the total weight of all nickel compounds processed. However, if your facility processes both the "parent" metal (nickel) as well as one or more nickel compounds, you must make threshold determinations for both nickel (CAS number 7440-02-0) and nickel compounds (chemical category code N495) because they are separately listed EPCRA Section 313 chemicals. If your facility exceeds thresholds for both the parent metal and compounds of that same metal, EPA allows you to file one combined report (e.g., one report for nickel compounds, including nickel) because the release information you will report in connection with metal category compounds will be the total pounds of the metal released. If you file one combined report, you should put the name of the metal compound category on the Form R. In the example above, the facility that exceeded reporting thresholds for both the nickel and nickel compounds chemical category could submit a single Form R for the nickel compounds chemical category, which would contain release and other waste management information for both nickel and nickel compounds. Do not put both names on the Form R.

The case of metal category compounds involving more than one metal should be noted. Some metal category compounds may contain more than one listed metal. For example, lead chromate is both a lead compound and a chromium compound. In such cases, if applicable thresholds are exceeded, you are required to file two separate reports, one for lead compounds and one for chromium compounds. Apply the total weight of the lead chromate to the threshold determinations for both lead compounds and chromium compounds. (Note: Only the quantity of each parent metal released or otherwise managed as waste, not the quantity of the compound, would be reported on the appropriate sections of both Form Rs. See B.5.)

Nitrate Compounds (water dissociable; reportable only when in aqueous solution)

For the category nitrate compounds (water dissociable; reportable only when in aqueous solution), the entire weight of the nitrate compound is counted in making threshold determinations. A nitrate compound is covered by this listing only when in water and only if dissociated. If no information is available on the identity of the type of nitrate that is manufactured, processed or otherwise used, assume that the nitrate compound exists as sodium nitrate.

B.4.e Threshold Determination for Persistent Bioaccumulative Toxic (PBT) Chemicals

There are two separate thresholds for EPCRA Section 313 PBT chemicals; these thresholds are set based on the chemicals' potential to persist and bioaccumulate in the environment. The manufacturing, processing and otherwise use thresholds for PBT chemicals is 100 pounds, while for the subset of PBTs chemicals that are highly persistent and highly bioaccumulative, it is 10 pounds. One exception is the dioxin and dioxin-like compounds chemical category. The threshold for this category is 0.1 gram. The PBT chemicals, their CAS numbers or chemical category code, and their reporting thresholds are listed in a table in the introductory section of B.4. See Table IIc of these instructions for lists of individual members of the dioxin and dioxin-like compounds chemical category and the polycyclic aromatic compounds (PACs) chemical category.

B.4.f. Mixtures and Other Trade Name Products

EPCRA Section 313 chemicals contained in mixtures and other trade name products must be factored into threshold determinations and release and other waste management calculations.

If your facility processed or otherwise used mixtures or other trade name products during the calendar year, you are required to use the best readily available data (or reasonable estimates if such data are not readily available) to determine whether the toxic chemicals in a mixture meet or exceed the *de*

minimis concentration and, therefore, whether they must be included in threshold determinations and release and other waste management calculations. If you know that a mixture or other trade name product contains a specific EPCRA Section 313 chemical, combine the amount of the EPCRA Section 313 chemical in the mixture or other trade name product with other amounts of the same EPCRA Section 313 chemical processed or otherwise used at your facility for threshold determinations and release and other waste management calculations. If you know that a mixture contains an EPCRA Section 313 chemical but it is present below the *de minimis* level, you do not have to consider the amount of the EPCRA Section 313 chemical present in that mixture for purposes of threshold determinations and release and other waste management calculations. PBT chemicals are not eligible for the *de minimis* exemption except lead when it is contained in stainless steel, brass or bronze alloy.

Observe the following guidelines in estimating concentrations of EPCRA Section 313 chemicals in mixtures when only limited information is available:

- If you only know the upper bound concentration, you must use it for threshold determinations (40 CFR Section372.30(b)(ii)).

- If you know the lower and upper bound concentrations of an EPCRA Section 313 chemical in a mixture, EPA recommends you use the midpoint of these two concentrations for threshold determinations.

- If you know only the lower bound concentration, EPA recommends you subtract out the percentages of any other known components to determine a reasonable upper bound concentration, and then determine a midpoint.

- If you have no information other than the lower bound concentration, EPA recommends you calculate a midpoint assuming an upper bound concentration of 100 percent.

B.5 Release and Other Waste Management Determinations for Metals, Metal Category Compounds, and Nitrate Compounds

Metal Category Compounds

Although the complete weight of the metal category compounds must be used in threshold determinations for the metal compounds category, only the weight of the metal portion of the metal category compound must be considered for release and other waste management determinations. Remember that for metal category compounds that consist of more than one metal, release and other waste management reporting must be based on the weight of each metal, provided that the appropriate thresholds have been exceeded.

Metals and Metal Category Compounds

For compounds within the metal compound categories, only the metal portion of the metal category compound must be considered in determining release and other waste management quantities for the metal category compounds. Therefore, if thresholds are separately exceeded for both the "parent" metal and its compounds, EPA allows you to file a combined Form R for the "parent" metal and its category compounds. This Form R would contain all of the release and other waste management information for both the "parent" metal and metal portion of the related metal category compounds. For example, you exceed thresholds for chromium. You also exceed thresholds for chromium compounds. Instead of filing two Form Rs you can file one combined Form R. This Form R would contain information on quantities of chromium released or otherwise managed as waste and the quantities of the chromium portion of the chromium compounds released or otherwise managed as waste. When filing one combined Form R for an EPCRA Section 313 metal and metal compound category, facilities should identify the chemical reported as the metal compound category name and code in Section 1 of the Form R. Note that this does not apply to the Form A. See Section E.7 in these instructions on the Form A. See Appendix B for more information about reporting the release and other waste management of metals and metal compounds.

Nitrate Compounds (water dissociable; reportable only in aqueous solution)

Although the complete weight of the nitrate compound must be used for threshold determinations for the nitrate compounds category only the nitrate portion of the compound should be used for release and other waste management calculations.

Example 9: Mixtures and Other Trade Name Products

Scenario #1: Your facility otherwise uses 12,000 pounds of an industrial solvent (Solvent X) for equipment cleaning. The Material Safety Data Sheet (MSDS) for the solvent indicates that it contains at least 50 percent n-hexane, an EPCRA Section 313 chemical; however, it also states that the solvent contains 20 percent non-hazardous surfactants. This is the only n-hexane-containing mixture used at the facility.

EPA recommends you follow these steps to determine if the quantity of the EPCRA Section 313 chemical in Solvent X exceeds the threshold for otherwise use.

1) Determine a reasonable maximum concentration for the EPCRA Section 313 chemical by subtracting out the non-hazardous surfactants (i.e., 100% - 20% = 80%).

2) Determine the midpoint between the known minimum (50%) and the reasonable maximum calculated above (i.e., (80% + 50%)/2 = 65%).

3) Multiply total weight of Solvent X otherwise used by 65% (0.65).

   12,000 pounds × 0.65 = 7,800 pounds

4) Because the total amount of n-hexane otherwise used at the facility was less than the 10,000-pound otherwise use threshold, the facility is not required to file a Form R for n-hexane.

Scenario #2: Your facility otherwise used 15,000 pounds of Solvent Y to clean printed circuit boards. The MSDS for the solvent lists only that Solvent Y contains at least 80 percent of an EPCRA Section 313 chemical that is only identified as chlorinated hydrocarbons.

EPA recommends you follow these steps to determine if the quantity of the EPCRA Section 313 chemical in the solvent exceeds the threshold for otherwise use.

1) Because the specific chemical is unknown, the Form R will be filed for "chlorinated hydrocarbons." This name will be entered into Part II, Section 2.1, "Mixture Component Identity." (Note: Because your supplier is claiming the EPCRA Section 313 chemical identity a trade secret, you do not have to file substantiation forms.)

2) The upper bound limit is assumed to be 100 percent and the lower bound limit is known to be 80 percent. Using this information, the specific concentration is estimated to be 90 percent (i.e., the mid-point between upper and lower limits).

   (100% + 80%)/2 = 90%

3) The total weight of Solvent Y is multiplied by 90 percent (0.90) when calculating for thresholds.

   15,000 × 0.90 = 13,500

4) Because the total amount of chlorinated hydrocarbons exceeds the 10,000-pound otherwise use threshold, you must file a Form R for this chemical.

How to Determine if Your Facility Must Submit a Form R or Is Eligible to Use Form A

Facility Name: _____ **Date Worksheet Prepared:** _____

EPCRA Section 313 Chemical or Chemical Category: _____ **Prepared By:** _____

CAS Registry Number: _____

Reporting Year: _____

Amounts of the EPCRA Section 313 chemical or chemical category manufactured, processed, or otherwise used.

Mixture Name or Other Identifier	Information Source	Total Weight (lb)	Percent EPCRA Section 313 Chemical by Weight	EPCRA Section 313 Chemical Weight (lb)	Amount of the EPCRA Section 313 Chemical or Chemical Category by Activity (lb):		
					Manufactured	Processed	Otherwise Used
1.							
2.							
3.							
4.							
Subtotal:					(A) _____ lb	(B) _____ lb	(C) _____ lb

Exempt quantity of the EPCRA Section 313 chemical or chemical category that should be excluded.

Mixture Name as Listed Above	Applicable Exemption (articles, facility, activity)	Fraction or Percent Exempt (if Applicable)	Amount of the EPCRA Section 313 Chemical Exempt from Above (lb):		
			Manufactured	Processed	Otherwise Used
1.					
2.					
3.					
4.					
Subtotal:			(A_1) _____ lb	(B_1) _____ lb	(C_1) _____ lb

Amount subject to threshold: ($A-A_1$) _____ lb ($B-B_1$) _____ lb ($C-C_1$) _____ lb

Compare to threshold for EPCRA Section 313 reporting. 25,000 lb 25,000 lb 10,000 lb

If any threshold is exceeded, reporting is required for all activities. Do not submit this worksheet with Form R or Form A; retain it for your records.

Figure 3A. EPCRA Section 313 Non-PBT Chemical Reporting Threshold Worksheet[1]

---

[1] Note: Chemicals listed as PBT have separate thresholds (dioxin and dioxin-like compounds chemical category = 0.1 g; highly persistent, highly bioaccumulative toxic chemicals = 10 lbs; all other PBT chemicals = 100 lbs). Make certain you are using the appropriate worksheet for the toxic chemical of concern.

How to Determine if Your Facility Must Submit a Form R or Is Eligible to Use Form A

Facility Name: _____   **Date Worksheet Prepared:** _____

EPCRA Section 313 Chemical or Chemical Category: _____   **Prepared By:** _____

CAS Registry Number: _____

Reporting Year: _____

Amounts of the EPCRA Section 313 chemical or chemical category manufactured, processed, or otherwise used.

Mixture Name or Other Identifier	Information Source	Total Weight (lb)	Percent EPCRA Section 313 Chemical by Weight	EPCRA Section 313 Chemical Weight (lb)	Amount of the EPCRA Section 313 Chemical or Chemical Category by Activity (lb):		
					Manufactured	Processed	Otherwise Used
1.							
2.							
3.							
4.							
Subtotal:					(A)_____ lb	(B)_____ lb	(C)_____ lb

Exempt quantity of the EPCRA Section 313 chemical or chemical category that should be excluded.

Mixture Name as Listed Above	Applicable Exemption (articles, facility, activity) [1]	Fraction or Percent Exempt (if Applicable)	Amount of the EPCRA Section 313 Chemical Exempt from Above (lb):		
			Manufactured	Processed	Otherwise Used
1.					
2.					
3.					
4.					
Subtotal:			(A₁)_____ lb	(B₁)_____ lb	(C₁)_____ lb

Amount subject to threshold: (A–A₁)_____ lb   (B–B₁)_____ lb   (C–C₁)_____ lb

Compare to threshold for EPCRA Section 313 reporting.   100 lb   100 lb   100 lb

If any threshold is exceeded, reporting is required for all activities. Do not submit this worksheet with Form R or Form A; retain it for your records.

Figure 3B. EPCRA Section 313 Reporting Threshold Worksheet for PBT Chemicals with 100 Pound Thresholds

---

[1] Note: Chemicals listed as PBT are not eligible for the de minimis exemption.

How to Determine if Your Facility Must Submit a Form R or Is Eligible to Use Form A

Facility Name: _____

EPCRA Section 313 Chemical or Chemical Category: _____

CAS Registry Number: _____

Reporting Year: _____

Date Worksheet Prepared: _____

Prepared By: _____

Amounts of the EPCRA Section 313 chemical or chemical category manufactured, processed, or otherwise used.

Mixture Name or Other Identifier	Information Source	Total Weight (lb)	Percent EPCRA Section 313 Chemical by Weight	EPCRA Section 313 Chemical Weight (lb)	Amount of the EPCRA Section 313 Chemical or Chemical Category by Activity (lb):		
					Manufactured	Processed	Otherwise Used
1.							
2.							
3.							
4.							
Subtotal:					(A)_____ lb	(B)_____ lb	(C)_____ lb

Exempt quantity of the EPCRA Section 313 chemical or chemical category that should be excluded.

Mixture Name as Listed Above	Applicable Exemption (articles, facility, activity) [1]	Fraction or Percent Exempt (if Applicable)	Amount of the EPCRA Section 313 Chemical Exempt from Above (lb):		
			Manufactured	Processed	Otherwise Used
1.					
2.					
3.					
4.					
Subtotal:			(A_1)_____ lb	(B_1)_____ lb	(C_1)_____ lb

Amount subject to threshold: $(A-A_1)$_____ lb $(B-B_1)$_____ lb $(C-C_1)$_____ lb

Compare to threshold for EPCRA Section 313 reporting. 10 lb 10 lb 10 lb

If any threshold is exceeded, reporting is required for all activities. Do not submit this worksheet with Form R or Form A; retain it for your records.

Figure 3C. EPCRA Section 313 Reporting Threshold Worksheet for PBT Chemicals with 10 Pound Threshold

[1] Note: Chemicals listed as PBT are not eligible for the de minimis exemption.

How to Determine if Your Facility Must Submit a Form R or Is Eligible to Use Form A

Facility Name: _____ **Date Worksheet Prepared:** _____

EPCRA Section 313 Chemical or Chemical Category: Dioxin and Dioxin-like Compounds **Prepared By:** _____

CAS Registry Number: _____

Reporting Year: _____

Amounts of the EPCRA Section 313 chemical or chemical category manufactured, processed, or otherwise used.

Mixture Name or Other Identifier	Information Source	Total Weight (g)	Percent EPCRA Section 313 Chemical by Weight	EPCRA Section 313 Chemical Weight (g)	Amount of the EPCRA Section 313 Chemical or Chemical Category by Activity (g):		
					Manufactured	Processed	Otherwise Used
1.							
2.							
3.							
4.							
Subtotal:					(A) _____ g	(B) _____ g	(C) _____ g

Exempt quantity of the EPCRA Section 313 chemical or chemical category that should be excluded.

Mixture Name as Listed Above	Applicable Exemption (articles, facility, activity) [1]	Fraction or Percent Exempt (if Applicable)	Amount of the EPCRA Section 313 Chemical Exempt from Above (g):		
			Manufactured	Processed	Otherwise Used
1.					
2.					
3.					
4.					
Subtotal:			(A₁) _____ g	(B₁) _____ g	(C₁) _____ g

Amount subject to threshold: (A–A₁) _____ g (B–B₁) _____ g (C–C₁) _____ g

Compare to threshold for EPCRA Section 313 reporting.   0.1 g   0.1 g   0.1 g

If any threshold is exceeded, reporting is required for all activities. Do not submit this worksheet with Form R or Form A; retain it for your records.

Figure 3D. EPCRA Section 313 Reporting Threshold Worksheet for Dioxin and Dioxin-Like Compounds Chemical Category

---

[1] Note: Chemicals listed as PBT are not eligible for the de minimis exemption.

C. Instructions for Completing TRI Form R

The following instructions provide information on how to enter data on a Form R. Some data entry fields of a Form R are done differently using the TRI's electronic reporting tool called TRI-MEweb than on a hard-copy paper form. TRI-MEweb also has automatic data quality tools that will assist users to enter valid data on their Form R and detect any errors on a Form R. This is why TRI-MEweb is EPA's preferred method to submit TRI data on a Form R.

Part I. Facility Identification Information

Section 1. Reporting Year

The reporting year is the calendar year to which the reported information applies, not the year in which you are submitting the report. Information for the 2012 reporting year must be submitted on or before July 1, 2013.

Section 2. Trade Secret Information

2.1 Are you claiming the EPCRA Section 313 chemical identified on Page 2 a trade secret?

Answer this question only after you have completed the rest of the report. The specific identity of the EPCRA Section 313 chemical being reported in Part II, Section 1 may be designated as a trade secret. If you are making a trade secret claim, mark "yes" and proceed to Section 2.2. Only check "yes" if you manufacture, process, or otherwise use the EPCRA Section 313 chemical whose identity is a trade secret. (See Section A.3 of these instructions for specific information on trade secrecy claims.) If you checked "no," proceed to Section 3; do not answer Section 2.2.

2.2 If "yes" in 2.1, is this copy sanitized or unsanitized?

Answer this question only after you have completed the rest of the report. Check "sanitized" if this copy of the report is the public version that does not contain the EPCRA Section 313 chemical identity but does contain a generic name that is structurally descriptive in its place, and if you have claimed the EPCRA Section 313 chemical identity trade secret

in Part I, Section 2.1. Otherwise, check "unsanitized."

Section 3. Certification

The certification statement must be signed by a senior official with management responsibility for the person (or persons) completing the form. A senior management official must certify the accuracy and completeness of the information reported on the form by signing and dating the certification statement. Each report must contain an original signature. You should print or type the name and title of the person who signs the statement in the space provided. This certification statement applies to all the information supplied on the form and should be signed only after the form has been completed.

Section 4. Facility Identification

4.1 Facility Name, Location, TRI Facility Identification Number and Tribal Country Name

Enter the full name that the facility presents to the public and its customers in doing business (e.g., the name that appears on invoices, signs, and other official business documents). Do not use a nickname for the facility (e.g., Main Street Plant) unless that is the legal name of the facility under which it does business. Also enter the physical street address, mailing address, city, county, three digit BIA code, if applicable, state, and ZIP code in the space provided. The street address provided must be the location where the EPCRA Section 313 chemicals are manufactured, processed, or otherwise used. You may not use PO Box as a facility address. If your mailing address and street address are the same, you should enter NA in the space for the mailing address. If the mailing address is outside of the US, include the FIPS country code, which may be found in Table IV.

If your facility is not in a county, put the name of your city, district (for example, District of Columbia), or parish (if you are in Louisiana) in the county block of the Form R and Form A as well as in the county field of TRI-MEweb. "NA" or "None" are not acceptable entries. TRI-MEweb provides a drop-down menu for the county name, including city districts and parish names.

If your facility is located on Indian country as defined by 18 USC §1151 you must enter the three digit Bureau of Indian Affairs (BIA) tribal code in the "City/County/Tribe/State/ZIP code" field. The BIA tribal codes are listed in Table V of the RFI. Facilities using TRI-MEweb to complete their forms will be asked if they are located within a tribe's Indian country and, upon answering "yes", be taken to a look up table to determine the correct BIA code.

If your facility is not located (overwhelming majority of TRI facilities are not in Indian Country) in Indian country as defined by 18 USC §1151 you must enter only the city, county (as applicable), state and zip code Facilities filing a paper form should leave blank in the BIA field, if not located within tribal boundaries. Facilities using TRI-MEweb to complete their forms will be required to check a specific checkbox if they are located within tribal lands and if they do not check that checkbox.

Location information for a facility that has previously submitted data to EPA.

If your facility has submitted a Form R in previous reporting years, and your facility is planning on submitting a TRI Form R, a TRI Facility Identification Number (TRIFID) has been already assigned to your facility. If you know your TRIFID, you should complete Form R Part I Section 4. If you do not know your TRI Facility Identification Number, you should visit the TRI Web page at: http://www.epa.gov/tri/contacts/contacts.htm for more information or contact your Regional TRI Program representative, or utilize Envirofacts on the Web to look up the address or facility name at: http://www.epa.gov/enviro.

Location information for a facility that has previously submitted data to EPA, but has changed physical location.

Hard-copy paper Form R: If your facility has moved do not enter your previously assigned TRI Facility Identification Number, enter "New Facility". If you are filing a separate Form R for each establishment at your facility, you should use the same "New Facility" field for each establishment. Utilize Envirofacts on the Web to look up the address or facility name at: http://www.epa.gov/enviro if you are uncertain if a TRIFID has been assigned to your new facility location.

TRI-MEweb: If your facility has moved, you will need to request that a new TRIFID be assigned to your facility, add a new facility. To request a new TRIFID, add a new facility account to TRI-MEweb and chose to report as a new reporting facility (option 3). TRI-MEweb will automatically generate a new TRIFID for your facility. The TRIFID assigned to your new reporting facility should be used in all future reporting of TRI data.

Location information for a facility that has changed ownership, but has not changed physical location.

The TRI Facility Identification Number (TRIFID) is established by the first Form R submitted by a facility at a particular location. Only a change in address warrants filing as a new facility; otherwise, the TRI Facility Identification Number is retained by the facility even if the facility changes name, ownership, production processes, NAICS codes, etc.

Hard-copy paper Form R: The TRIFID identification number will always stay with the physical location of a facility. If a new facility unit moves to this location it should use this TRIFID. Establishments of a facility (for facilities that report by part) that report separately should use the TRIFID of the primary facility.

TRI-MEweb: If your facility has changed ownership during the reporting year but not its physical location, the facility does not require a new TRIFID. Use the TRIFID assigned to previous owner. TRI-MEweb can be used to update facility information due to change of ownership.

Location reporting TRI releases for the first time to EPA.

Hard-copy paper Form R: If you are preparing a hard-copy TRI form for the first time for your facility's location and have never reported to TRI in prior years, you should enter "New Facility" in the space on the hardcopy form designated for the TRI Facility Identification number (TRIFID).

TRI-MEweb: If your facility is reporting for the first time, upon creating your CDX account, and adding TRI-MEweb application, you will be prompted to add a new facility account into TRI-MEweb. TRI-MEweb will automatically generate a new TRIFID for your facility. The TRIFID assigned to your new reporting facility should be used in all future reporting of TRI data.

4.2 Full or Partial Facility Indication and Federal Facility Designation

EPCRA Section 313 requires reports by "facilities," which are defined as "all buildings, equipment, structures, and other stationary items which are located on a single site or on contiguous or adjacent sites and which are owned or operated by the same person (or by any person which controls, is controlled by, or under common control with such person). A facility may contain more than one establishment."

EPCRA Section 313 defines establishment as "an economic unit, generally at a single physical location, where business is conducted or where services or industrial operations are performed." Under Section 372.30(c) of the reporting rule, you may submit a separate Form R for each establishment or for groups of establishments in your facility, provided all releases and other waste management activities and source reduction activities involving the EPCRA Section 313 chemical from the entire facility are reported. This allows you the option of reporting separately on the activities involving an EPCRA Section 313 chemical at each establishment, or group of establishments (e.g., part of a covered facility), rather than submitting a single Form R for that EPCRA Section 313 chemical for the entire facility. However, if an establishment or group of establishments does not manufacture, process, or otherwise use or release or otherwise manage as waste an EPCRA Section 313 chemical, you do not have to submit a report for that establishment or group of establishments for that particular chemical. (See also Section B.2.b of these instructions.)

A covered facility must report all releases and other waste management activities and source reduction activities of an EPCRA Section 313 chemical if the facility meets a reporting threshold for that EPCRA Section 313 chemical. Whether submitting a report for the entire facility or separate reports for the establishments, the threshold determination must be made based on the entire facility. Indicate in Section 4.2 whether your report is for the entire covered facility as a whole or for part of a covered facility (i.e., one or more establishments).

Federal facilities and contractors at federal facilities (GOCOs: Government-owned, contractor-operated facilities) should check either 4.2c or 4.2d, but not both. Federal facilities should check 4.2c, even if their TRI reports contain release and other waste management information from contractors located at the facility. Contractors at federal facilities, which are required by EPCRA Section 313 to file TRI reports independently of the federal facility, should check 4.2d. This information is important to prevent duplication of federal facility data. (See Appendix A for further guidance on these instructions.)

4.3 Technical Contact

Enter the name and telephone number (including area code) of a technical representative whom EPA, state, or tribal officials may contact for clarification of the information reported on Form R. You should also enter an email address for this person. EPA encourages facilities to provide an email address for the Technical Contact on their TRI submissions because they will be able to receive important program updates and email alerts notifying them when their FDP has been updated and published for their review. If the technical contact does not have an email address you should enter NA. This contact person does not have to be the same person who prepares the report or signs the certification statement and does not necessarily need to be someone at the location of the reporting facility. However, this person should be familiar with the details of the report so that he or she can answer questions about the information provided.

4.4 Public Contact

Enter the name and telephone number (including area code) of a person who can respond to questions from the public about the form. You should also enter an e-mail address for this person. If you choose to designate the same person as both the Technical and the Public Contact, or you do not have a Public Contact, you may enter "Same as Section 4.3" in this space. This contact person does not have to be the same person who prepares the form or signs the Certification Statement and does not necessarily need to be someone at the location of the reporting facility.

4.5 North American Industry Classification System (NAICS) Codes

The North American Industry Classification System (NAICS) is the economic classification system that replaces the 1987 SIC code system for TRI Reporting beginning with the RY 2006 EPCRA Section 313 reporting (71 FR 32464). Enter the appropriate six digit North American Industry Classification System (NAICS) Code that is the primary NAICS code for your facility in Section 4.5(a). For Reporting Year 2008 and beyond, use 2007 NAICS codes. Enter any other applicable NAICS for your facility in 4.5 (b)-(f). Table I lists the TRI-covered NAICS codes. If you do not know your NAICS code, consult the 2007 NAICS Manual (see Section B.2 of these instructions for ordering information) or check the SIC to NAICS crosswalk tables at: http://www.census.gov.

4.6   Dun & Bradstreet Number(s)

Enter the nine digit number assigned by Dun & Bradstreet (D&B) for your facility or each establishment within your facility. These numbers code the facility for financial purposes. This number may be available from your facility's treasurer or financial officer. You can also obtain the numbers from Dun & Bradstreet by calling 1-888-814-1435, or by visiting this website: https://www.dnb.com/product/dlw/form_cc4.htm. If a facility does not subscribe to the D&B service, a number can be obtained, toll free at 800 234-3867 (8:00 AM to 6:00 PM, EST) or on the Web at: http://www.dnb.com. If none of your establishments has been assigned a D&B number, you should enter NA in box (a). If only some of your establishments have been assigned D&B numbers, enter those numbers in Part I, section 4.6.

Section 5.   Parent Company Information

You must provide information on your parent company. For TRI Reporting purposes, your parent company is the highest level U.S. company which directly owns at least 50 percent of the voting stock of your company. If there is no higher level U.S. company, select the "No U.S. Parent Company parent (for TRI reporting purposes)" check box. Corporate names should be treated as parent company names for companies with multiple facility sites. For example, the Bestchem Corporation is not owned or controlled by any other corporation but has sites throughout the country whose names begin with Bestchem. In this case, Bestchem Corporation should be listed as the parent company. Note that a facility that is a 50:50 joint venture is its own parent company. When a facility is owned by more than one company and none of the facility owners directly owns at least 50 percent of its voting stock, the facility should provide the name of the parent company of either the facility operator or the owner with the largest ownership interest in the facility.

5.1   Name of Parent Company

Enter the name of the corporation or other business entity that is your highest level U.S. parent company. If your facility has no higher level U.S. company, select the "No U.S. Parent Company (for TRI reporting purposes)" check box.

5.2   Parent Company's Dun & Bradstreet Number

Enter the D&B number for your ultimate U.S. parent company, if applicable. The number may be obtained from the treasurer or financial officer of the company or by calling 1-888-814-1435, or by visiting this website: https://www.dnb.com/product/dlw/form_cc4.htm. If your parent company does not have a D&B number, you should check the NA box.

Part II. Chemical Specific Information

In Part II, you are to report on:

- The EPCRA Section 313 chemical being reported;

- The general uses and activities involving the EPCRA Section 313 chemical at your facility;

- On-site releases of the EPCRA Section 313 chemical from the facility to air, water, and land;

- Quantities of the EPCRA Section 313 chemical transferred to off-site locations;

- Information for on-site and off-site disposal, treatment, energy recovery, and recycling of the EPCRA Section 313 chemical; and

- Source reduction activities.

Section 1. EPCRA Section 313 Chemical Identity

1.1 CAS Number

Enter the Chemical Abstracts Service (CAS) registry number in Section 1.1 exactly as it appears in Table II of these instructions for the chemical being reported. CAS numbers are cross-referenced with an alphabetical list of chemical names in Table II. If you are reporting one of the EPCRA Section 313 chemical categories (e.g., chromium compounds), you should enter the applicable category code in the CAS number space. EPCRA Section 313 chemical category codes are listed below and can also be found in Table IIc and Appendix B.

EPCRA Section 313 Chemical Category Codes:

N010	Antimony compounds
N020	Arsenic compounds
N040	Barium compounds
N050	Beryllium compounds
N078	Cadmium compounds
N084	Chlorophenols
N090	Chromium compounds
N096	Cobalt compounds
N100	Copper compounds
N106	Cyanide compounds
N120	Diisocyanates
N150	Dioxin and dioxin-like compounds
N171	Ethylenebisdithiocarbamic acid, salts and esters (EBDCs)

N230	Certain glycol ethers
N420	Lead compounds
N450	Manganese compounds
N458	Mercury compounds
N495	Nickel compounds
N503	Nicotine and salts
N511	Nitrate compounds (water dissociable, reportable only in aqueous solution)
N575	Polybrominated biphenyls (PBBs)
N583	Polychlorinated alkanes (C10 to C13)
N590	Polycyclic aromatic compounds (PACs)
N725	Selenium compounds
N740	Silver compounds
N746	Strychnine and salts
N760	Thallium compounds
N770	Vanadium compounds
N874	Warfarin and salts
N982	Zinc compounds

If you are making a trade secret claim, you must report the CAS number or category code on your unsanitized Form R and unsanitized substantiation form. Do not include the CAS number or category code on your sanitized Form R or sanitized substantiation form.

1.2 EPCRA Section 313 Chemical or Chemical Category Name

Enter the name of the EPCRA Section 313 chemical or chemical category exactly as it appears in Table II. If the EPCRA Section 313 chemical name is followed by a synonym in parentheses, report the chemical by the name that directly follows the CAS number (i.e., not the synonym). If the EPCRA Section 313 chemical identity is actually a product trade name (e.g., Dicofol), the *Chemical Abstracts 9th Collective Index* name is listed below it in brackets. You may report either name in this case.

Do not list the name of a chemical that does not appear in Table II, such as individual members of an EPCRA Section 313 chemical category. For example, if you use silver chloride, **do not** report silver chloride with its CAS number. Report this chemical as "silver compounds" with its category code, N740.

If you are making a trade secret claim, you must report the specific EPCRA Section 313 chemical identity on your unsanitized Form R and unsanitized substantiation form. Do not report the name of the EPCRA Section 313 chemical on your sanitized Form R or sanitized substantiation form. Include a generic name that is structurally descriptive in Part II, Section 1.3 of your sanitized Form R report.

EPA requests that the EPCRA Section 313 chemical, chemical category, or generic name also be placed in the box marked "Toxic Chemical, Category, or Generic Name" in the upper right-hand corner on all pages of Form R. While this space is not a required data element, providing this information will help you in preparing a complete Form R report.

1.3   Generic Chemical Name

Complete Section 1.3 only if you are claiming the specific EPCRA Section 313 chemical identity of the EPCRA Section 313 chemical as a trade secret and have marked the trade secret block in Part I, Section 2.1 on Page 1 of Form R. Enter a generic chemical name that is descriptive of the chemical structure. You should limit the generic name to seventy characters (e.g., numbers, letters, spaces, punctuation) or less. Do not enter mixture names in Section 1.3; see Section 2 below.

In-house plant codes and other substitute names that are not structurally descriptive of the EPCRA Section 313 chemical identity being withheld as a trade secret are not acceptable as a generic name. The generic name must appear on both sanitized and unsanitized Form Rs, and the name must be the same as that used on your substantiation forms.

Section 2.   Mixture Component Identity

Do not complete this section if you have completed Section 1 of Part II. Report the generic name provided to you by your supplier in this section if your supplier is claiming the chemical identity proprietary or trade secret. Do not answer "yes" in Part I, Section 2.1 on Page 1 of the form if you complete this section. You do not need to supply trade secret substantiation forms for this EPCRA Section 313 chemical because it is your supplier who is claiming the chemical identity a trade secret.

Example 10: Mixture Containing Unidentified EPCRA Section 313 Chemical

Your facility uses 20,000 pounds of a solvent that your supplier has told you contains 80 percent "chlorinated aromatic," their generic name for a non-PBT EPCRA Section 313 chemical subject to reporting under EPCRA Section 313. You, therefore, have used 16,000 pounds of some EPCRA Section 313 chemical and that exceeds the "otherwise use" threshold for a non-PBT chemical. You would file a Form R and enter the name "chlorinated aromatic" in the space provided in Part II, Section 2.

2.1   Generic Chemical Name Provided by Supplier

Enter the generic chemical name in this section only if the following three conditions apply:

1.) You determine that the mixture contains an EPCRA Section 313 chemical but the only identity you have for that chemical is a generic name;

2.) You know either the specific concentration of that EPCRA Section 313 chemical component or a maximum or average concentration level; and

3.) You multiply the concentration level by the total annual amount of the whole mixture processed or otherwise used and determine that you meet the process or otherwise use threshold for that single, generically identified mixture component.

Section 3.   Activities and Uses of the EPCRA Section 313 Chemical at the Facility

Indicate whether the EPCRA Section 313 chemical is manufactured (including imported), processed, or otherwise used at the facility and the general nature of such activities and uses at the facility during the calendar year (see Figure 4). You are not required to report on Form R the quantity manufactured, processed or otherwise used. Report activities that take place only at your facility, not activities that take place at other facilities involving your products. You must check all the boxes in this section that apply. Refer to the definitions of "manufacture," "process," and "otherwise use" in Section B.3.a or

Part 40, Section 372.3 of the CFR for additional explanations.

3.1 Manufacture the EPCRA Section 313 Chemical

Persons who manufacture (including import) the EPCRA Section 313 chemical must check at least one of the following:

a. *Produce* — The EPCRA Section 313 chemical is produced at the facility.

b. *Import* — The EPCRA Section 313 chemical is imported by the facility into the Customs Territory of the United States. (See Section B.3.a of these instructions for further clarification of import.)

And check at least one of the following:

c. *For on-site use/processing* — The EPCRA Section 313 chemical is produced or imported and then further processed or otherwise used at the same facility. If you check this block, generally you should also check at least one item in Part II, Section 3.2 or 3.3.

d. *For sale/distribution* — The EPCRA Section 313 chemical is produced or imported specifically for sale or distribution outside the manufacturing facility.

e. *As a byproduct* — The EPCRA Section 313 chemical is produced coincidentally during the manufacture, processing, or otherwise use of another chemical substance or mixture and, following its production, is separated from that other chemical substance or mixture. EPCRA Section 313 chemicals produced as a result of waste management are also considered byproducts.

f. *As an impurity* — The EPCRA Section 313 chemical is produced coincidentally as a result of the manufacture, processing, or otherwise use of another chemical but is not separated and remains in the mixture or other trade name product with that other chemical.

In summary, if you are a manufacturer of the EPCRA Section 313 chemical, you must check (a) and/or (b), and at least one of (c), (d), (e), and (f) in Section 3.1.

3.2 Process the EPCRA Section 313 Chemical

Persons who process the EPCRA Section 313 chemical must check at least one of the following:

a. *As a reactant* — A natural or synthetic EPCRA Section 313 chemical is used in chemical reactions for the manufacture of another chemical substance or of a product. Includes but is not limited to, feedstocks, raw materials, intermediates, and initiators.

b. *As a formulation component* — An EPCRA Section 313 chemical is added to a product (or product mixture) prior to further distribution of the product that acts as a performance enhancer during use of the product. Examples of EPCRA Section 313 chemicals used in this capacity include, but are not limited to, additives, dyes, reaction diluents, initiators, solvents, inhibitors, emulsifiers, surfactants, lubricants, flame retardants, and rheological modifiers.

c. *As an article component* — An EPCRA Section 313 chemical becomes an integral component of an article distributed for industrial, trade, or consumer use. One example is the pigment components of paint applied to a chair that is sold.

d. *Repackaging* — This consists of processing or preparation of an EPCRA Section 313 chemical (or product mixture) for distribution in commerce in a different form, state, or quantity. This includes, but is not limited to, the transfer of material from a bulk container, such as a tank truck to smaller containers such as cans or bottles.

e. *As an impurity* — The EPCRA Section 313 chemical is processed but is not separated and remains in the mixture or other trade name product with that/those other chemical(s).

3.3 Otherwise Use the EPCRA Section 313 Chemical (non-incorporative activities)

Persons who otherwise use the EPCRA Section 313 chemical must check at least one of the following:

a. *As a chemical processing aid* — An EPCRA Section 313 chemical that is added to a reaction mixture to aid in the manufacture or synthesis of another chemical substance but is not intended to remain in or become part of the product or product mixture is otherwise used as chemical processing aid. Examples of such EPCRA Section 313 chemicals include, but are not limited to, process solvents, catalysts, inhibitors, initiators, reaction terminators, and solution buffers.

b. *As a manufacturing aid* — An EPCRA Section 313 chemical that aids the manufacturing process but does not become part of the resulting product and is not added to the reaction mixture during the manufacture or synthesis of another chemical substance is otherwise used as a manufacturing aid. Examples include, but are not limited to,

process lubricants, metalworking fluids, coolants, refrigerants, and hydraulic fluids.

c. *Ancillary or other use* — An EPCRA Section 313 chemical that is used at a facility for purposes other than aiding chemical processing or manufacturing as described above is otherwise used as an ancillary or other use. Examples include, but are not limited to, cleaners, degreasers, lubricants, fuels, EPCRA Section 313 chemicals used for treating wastes, and EPCRA Section 313 chemicals used to treat water at the facility.

SECTION 1. TOXIC CHEMICAL IDENTITY
(Important: DO NOT complete this section if you are reporting a mixture component in Section 2 below.)

1.1	CAS Number (Important: Enter only one number exactly as it appears on the Section 313 list. Enter category code if reporting a chemical category.)
	334-88-3

1.2	Toxic Chemical or Chemical Category Name (Important: Enter only one name exactly as it appears on the Section 313 list.)
	Diazomethane

1.3	Generic Chemical Name (Important: Complete only if Part I, Section 2.1 is checked "Yes". Generic Name must be structurally descriptive.)

SECTION 2. MIXTURE COMPONENT IDENTITY	(Important: DO NOT complete this section if you completed Section 1.)

2.1	Generic Chemical Name Provided by Supplier (Important: Maximum of 70 characters, including numbers, letters, spaces, and punctuation.)

SECTION 3. ACTIVITIES AND USES OF THE TOXIC CHEMICAL AT THE FACILITY
(Important: Check all that apply.)

3.1	Manufacture the toxic chemical:	3.2	Process the toxic chemical:	3.3	Otherwise use the toxic chemical:
	a. ☑ Produce b. ☐ Import	a. ☑ As a reactant		a. ☐ As a chemical processing aid	
	If Produce or Import	b. ☐ As a formulation component		b. ☐ As a manufacturing aid	
	c. ☑ For on-site use/processing	c. ☐ As an article component		c. ☐ Ancillary or other use	
	d. ☑ For sale/distribution	d. ☐ Repackaging			
	e. ☐ As a byproduct	e. ☐ As an impurity			
	f. ☐ As an impurity				

Figure 4. Reporting EPCRA Section 313 Chemicals

Section 4. Maximum Amount of the EPCRA Section 313 Chemical On-site at Any Time during the Calendar Year

For data element 4.1 of Part II, insert the code (see codes below) that indicates the maximum quantity of the EPCRA Section 313 chemical (e.g., in storage tanks, process vessels, on-site shipping containers, or in wastes generated) at your facility at any time during the calendar year. If the EPCRA Section 313 chemical was present at several locations within your facility, use the maximum total amount present at the entire facility at any one time. While range reporting is not allowed for PBT chemicals elsewhere on the Form R, range reporting for PBT chemicals is allowed for the Maximum Amount On-site.

Example 11: Manufacturing and Processing Activities of EPCRA Section 313 Chemicals

In the two examples below, it is assumed that the threshold quantities for manufacture, process, or otherwise use (25,000 pounds, 25,000 pounds, and 10,000 pounds, respectively for non-PBT chemicals; 100 pounds for certain PBT chemicals; 10 pounds for highly persistent, highly bioaccumulative toxic chemicals; and 0.1 grams for the PBT chemical category comprised of dioxin and dioxin-like compounds) have been exceeded and the reporting of EPCRA Section 313 chemicals is therefore required.

1. Your facility manufactures diazomethane. Fifty percent is sold as a product, thus it is processed. The remaining fifty percent is reacted with alpha-naphthylamine, forming N-methyl-alpha-naphthylamine and also producing nitrogen gas.

- Your company manufactures diazomethane, an EPCRA Section 313 chemical, both for sale/ distribution as a commercial product and for on-site use/processing as a feedstock in the N-methyl-alpha-naphthylamine production process. Because the diazomethane is a reactant, it is also processed. See Figure 4 for how this information would be reported in Part II, Section 3 of Form R.

- Your facility also processes alpha-naphthylamine, as a reactant to produce N-methyl-alpha-naphthylamine, a chemical not on the EPCRA Section 313 list.

2. Your facility is a commercial distributor of Missouri bituminous coal, which contains mercury at 1.5 ppm (w:w). You should check the box on the Form R at Part II, Section 3.2.e for processing mercury as an impurity.

Weight Range in Pounds

Range Code	From...	To...
01	0	99
02	100	999
03	1,000	9,999
04	10,000	99,999
05	100,000	999,999
06	1,000,000	9,999,999
07	10,000,000	49,999,999
08	50,000,000	99,999,999
09	100,000,000	499,999,999
10	500,000,000	999,999,999
11	1 billion	more than 1 billion

If the EPCRA Section 313 chemical present at your facility was part of a mixture or other trade name product, determine the maximum quantity of the EPCRA Section 313 chemical present at the facility by calculating the weight percent of the EPCRA Section 313 chemical only.

Do not include the weight of the entire mixture or other trade name product. These data may be found in the Tier II form your facility may have prepared under Section 312 of EPCRA. See Part 40, Section 372.30(b) of the CFR for further information on how to calculate the weight of the EPCRA Section 313 chemical in the mixture or other trade name product. For EPCRA Section 313 chemical categories (e.g., nickel compounds), include all chemical compounds in the category when calculating the maximum amount, using the entire weight of each compound.

Weight Range in Grams (Dioxin and Dioxin-like Compounds)

When reporting for the dioxin and dioxin-like compounds category use the following gram quantity range codes:

Range Code	From...	To...
12	0	0.099
13	0.1	0.99
14	1.0	9.99
15	10	99
16	100	999
17	1,000	9,999
18	10,000	99,999
19	100,000	999,999
20	1,000,000	more than 1 million

Section 5. Quantity of the Toxic Chemical Entering Each Environmental Medium On-site

In Section 5, you must account for the total aggregate on-site releases of the EPCRA Section 313 chemical to the environment from your facility for the calendar year.

On-site releases to the environment include emissions to the air, discharges to surface waters, and releases to land and underground injection wells.

For all toxic chemicals (except the dioxin and dioxin-like compound category), do not enter the values in Section 5 in gallons, tons, liters, or any measure other than pounds. You must also enter the values as whole numbers (do not use scientific notation). Numbers following a decimal point are not acceptable for toxic chemicals other than those designated as PBT chemicals. For PBT chemicals, facilities should report release and other waste management quantities greater than 0.1 pound (except the dioxin and dioxin-like compounds category), provided the accuracy and the underlying data on which the estimate is based supports this level of precision.

For the dioxin and dioxin-like compounds category, facilities should report at a level of precision supported by the accuracy of the underlying data and the estimation techniques on which the estimate is based. For the dioxin and dioxin like compounds chemical category, which has a reporting threshold of 0.1 gram, facilities need only report all release and other waste management quantities greater than 100 micrograms (i.e., 0.0001 grams). (See Example 12) Notwithstanding the numeric precision used when determining reporting eligibility thresholds, facilities should report on Form R to the level of accuracy that their data supports, up to seven digits to the right of the decimal. EPA's reporting software and data management systems support data precision up to seven digits to the right of the decimal.

Example 12: Reporting Dioxins and Dioxin-Like Compounds

If the total quantity for Section 5.2 of the Form R (i.e., stack or point air emissions) is 0.00005 grams or less, then zero can be entered. If the total quantity is between 0.00005 and 0.0001 grams, then 0.0001 grams can be entered or the actual number can be entered (e.g., 0.000075).

NA vs. a Numeric Value (e.g., Zero). Generally, NA is applicable if the waste stream that contains or contained the EPCRA Section 313 chemical is not directed to the relevant environmental medium, or if leaks, spills and fugitive emissions cannot occur. If the waste stream that contains or contained the EPCRA Section 313 chemical is directed to the environmental medium, or if leaks, spills or fugitive emissions can occur, NA should not be used, even if treatment or emission controls result in a release of zero. If the annual aggregate release of that chemical was equal to or less than 0.5 pound, the value reported is zero (unless the chemical is a listed PBT chemical).

For Section 5.1, NA generally is not applicable for volatile organic compounds (VOCs). For Section 5.5.4, NA generally would not be applicable, recognizing the possibility of accidental spills or leaks of the EPCRA Section 313 chemical.

An example that illustrates the use of NA vs. a numeric value (e.g., zero) would be nitric acid involved in a facility's processing activities. If the facility neutralizes the wastes containing nitric acid to a pH of 6 or above, then the facility reports a release of zero for the EPCRA Section 313 chemical, not NA. Another example is when the facility has no underground injection well, in which case NA should be entered in Part I, Section 4.10 and checked in Part II, Section 5.4.1 and 5.4.2 of Form R. Also, if the facility does not landfill the acidic waste, NA should be checked in Part II, Section 5.5.1.B of Form R.

All releases of the EPCRA Section 313 chemical to the air must be classified as either stack or fugitive emissions, and included in the total quantity reported for these releases in Sections 5.1 and 5.2. Instructions for columns A, B, and C follow the discussions of Sections 5.1 through 5.5.

5.1 Fugitive or Non-Point Air Emissions

Report the total of all releases of the EPCRA Section 313 chemical to the air that are not released through stacks, vents, ducts, pipes, or any other confined air stream. You must include (1) fugitive equipment leaks from valves, pump seals, flanges, compressors, sampling connections, open-ended lines, etc.; (2) evaporative losses from surface impoundments and spills; (3) releases from building ventilation systems; and (4) any other fugitive or non-point air emissions. Engineering estimates and mass balance calculations (using purchase records, inventories, engineering knowledge or process specifications of the quantity of the EPCRA Section 313 chemical entering product, hazardous waste manifests, or monitoring records) may be useful in estimating fugitive emissions. You should check the NA box in Section 5.1 if you do not engage in activities that result in fugitive or non-point air emissions of this listed toxic chemical. For VOCs, NA generally would not be applicable.

5.2 Stack or Point Air Emissions

Report the total of all releases of the EPCRA Section 313 chemical to the air that occur through stacks, confined vents, ducts, pipes, or other confined air streams. You must include storage tank emissions. Air releases from air pollution control equipment would generally fall in this category. Monitoring data, engineering estimates, and mass balance calculations may help you to complete this section. You should check the NA box in Section 5.2 if there are no stack air activities involving the waste stream that contains or contained the EPCRA Section 313 chemical.

5.3 Discharges to Receiving Streams or Water Bodies

In Section 5.3 you are to enter all the names of the streams or water bodies to which your facility directly discharges the EPCRA Section 313 chemical on which you are reporting. A total of three spaces is provided on Page 2 of Form R. If you discharge the EPCRA Section 313 chemical to more than three streams or water bodies, you should photocopy Page 2 of Form R as many times as necessary and then number the boxes consecutively for each stream or water body. At the bottom of Page 2 you will find instructions for indicating the total number of Page 2s that you are submitting as part of the Form R as well as indicating the sequence of those pages. Enter the name of each receiving stream or surface water body to which the EPCRA Section 313 chemical being reported is directly discharged. Report the name of the receiving stream or water body as it appears on the permit for the facility. If the stream is not included in the NPDES permit or its name is not identified in the NPDES permit, enter the name of the off-site stream or water body by which it is publicly known or enter the first publicly named water body to which the receiving waters are a tributary, if the receiving waters are unnamed. Do not list a series of streams through which the EPCRA Section 313 chemical flows. Be sure to include all the receiving streams or water bodies that receive stormwater runoff from your facility. Do not enter names of streams to which off-site treatment plants discharge. You should check the NA box in section 5.3 if there are no discharges to receiving streams or water bodies of the waste stream that contains or contained the EPCRA Section 313 chemical (See discussion of NA vs. a Numeric Value (e.g., Zero) in the introduction of Section 5).

Enter the total annual amount of the EPCRA Section 313 chemical released from all discharge points at the facility to each receiving stream or water body. Include process outfalls such as pipes and open trenches, releases from on-site wastewater treatment systems, and the contribution from stormwater runoff, if applicable (see instructions for column C below). Do not include discharges to a POTW or other off-site wastewater treatment facilities in this section. These off-site transfers must be reported in Part II, Section 6 of Form R. Wastewater analyses and flowmeter data may provide the quantities you will need to complete this section.

Discharges of listed acids (e.g., hydrogen fluoride, nitric acid) may be reported as zero if the discharges have been neutralized to pH 6 or above. If wastewater containing a listed acid is discharged below pH 6, then releases of the acid must be reported. In this case, pH measurements may be used to estimate the amount of mineral acid released.

5.4.1 Underground Injection On-Site to Class I Wells

Enter the total amount of the EPCRA Section 313 chemical that was injected into Class I wells at the facility. Chemical analyses, injection rate meters, and RCRA Hazardous Waste Generator Reports are

good sources for obtaining data that will be useful in completing this section. You should check the NA box in Section 5.4.1 if you do not inject the waste stream that contains or contained the EPCRA Section 313 chemical into Class I underground wells (See discussion of NA vs. a Numeric Value (e.g., Zero) in the introduction of Section 5).

5.4.2 Underground Injection On-site to Class II-V Wells

Enter the total amount of the EPCRA Section 313 chemical that was injected into wells at the facility other than Class I wells. Chemical analyses and injection rate meters are good sources for obtaining data that will be useful in completing this section. You should check the NA box in Section 5.4.2 if you do not inject the waste stream that contains or contained the EPCRA Section 313 chemical into Class II-V underground wells (See discussion of NA vs. a Numeric Value (e.g., Zero) in the introduction of Section 5).

5.5 Disposal to Land On-site

Six predefined subcategories for reporting quantities released to land within the boundaries of the facility are provided. Do not report land disposal at off-site locations in this section. Accident histories and spill records may be useful (e.g., release notification reports required under Section 304 of EPCRA, Section 103 of CERCLA, and accident histories required under Section112(r)(7)(B)(ii) of the Clean Air Act). Where relevant, you should check the NA box in sections 5.5.1A through 5.5.3 if there are no disposal activities for the waste stream that contains or contained the EPCRA Section 313 chemical (See discussion of NA vs. a Numeric Value (e.g., Zero) in the introduction of Section 5). For 5.5.4, facilities generally should report zero, recognizing the potential for spills or leaks.

5.5.1A RCRA Subtitle C Landfills

Enter the total amount of the EPCRA Section 313 chemical that was placed in RCRA Subtitle C landfills. EPA has not required facilities to estimate leaks from landfills because the amount of the EPCRA Section 313 chemical has already been reported as a release.

5.5.1B Other Landfills

Enter the total amount of the EPCRA Section 313 chemical that was placed in landfills other than RCRA Subtitle C landfills. EPA has not required facilities to estimate leaks from landfills because the amount of the EPCRA Section 313 chemical has already been reported as a release.

5.5.2   Land Treatment/Application Farming

Land treatment is a disposal method in which a waste containing an EPCRA Section 313 chemical is applied onto or incorporated into soil. While this disposal method is considered a release to land, any volatilization of EPCRA Section 313 chemicals into the air occurring during the disposal operation must not be included in this section but must be included in the total fugitive air releases reported in Part II, Section 5.1 of Form R.

Surface Impoundments

A surface impoundment is a natural topographic depression, man-made excavation, or diked area formed primarily of earthen materials (although some may be lined with man-made materials), that is designed to hold an accumulation of liquid wastes or wastes containing free liquids. Examples of surface impoundments are holding, settling, storage, and elevation pits; ponds, and lagoons. If the pit, pond, or lagoon is intended for storage or holding without discharge, it would be considered to be a surface impoundment used as a final disposal method. A facility must determine, to the best of its ability, the percentage of a volatile chemical, e.g., benzene, that is in waste sent to a surface impoundment that evaporates during the reporting year. The facility must report this as a fugitive air emission in section 5.1. The balance should be reported in either section 5.5.3A or 5.5.3B.

Quantities of the EPCRA Section 313 chemical released to surface impoundments that are used merely as part of a wastewater treatment process generally should not be reported in this section. However, if an impoundment accumulates sludges containing the EPCRA Section 313 chemical, you must include an estimate in this section unless the sludges are removed and otherwise disposed of (in which case they must be reported under the appropriate section of the form). For the purposes of this reporting, storage tanks are not considered to be a type of disposal and are not to be reported in this section of Form R.

5.5.3A   RCRA Subtitle C Surface Impoundments

Enter the total amount of the EPCRA Section 313 chemical that was placed in RCRA Subtitle C surface impoundments.

5.5.3B   Other Surface Impoundments

Enter the total amount of the EPCRA Section 313 chemical that was placed in surface impoundments other than RCRA Subtitle C surface impoundments.

5.5.4   Other Disposal

Includes any amount of an EPCRA Section 313 chemical released to land that does not fit the categories of landfills, land treatment, or surface impoundment. This other disposal would include any spills or leaks of EPCRA Section 313 chemicals to land. For example, 2,000 pounds of benzene leaks from an underground pipeline into the land at a facility. Because the pipe was only a few feet from the surface at the erupt point, 30 percent of the benzene evaporates into the air. The 600 pounds released to the air would be reported as a fugitive air release (Part II, Section 5.1) and the remaining 1,400 pounds would be reported as a release to land, other disposal (Part II, Section 5.5.4).

Section 5 Column A: Total Release

Only on-site releases of the EPCRA Section 313 chemical to the environment for the calendar year are to be reported in this section of Form R. The total on-site releases from your facility do not include transfers or shipments of the EPCRA Section 313 chemical from your facility for sale or distribution in commerce, or of wastes to other facilities for disposal, treatment, energy recovery, or recycling (see Part II, Section 6 of these Instructions). Both routine releases, such as fugitive air emissions, and accidental or non-routine releases, such as chemical spills, must be included in your estimate of the quantity released.

Releases of Less Than 1,000 Pounds. For total annual releases or off-site transfers of an EPCRA Section 313 chemical from the facility of less than 1,000 pounds, the amount may be reported either as an estimate or by using the range codes that have been developed (range reporting in section 5 does not apply to PBT chemicals). The reporting range codes to be used are:

Code	Range (pounds)
A	1-10
B	11-499
C	500-999

Do not enter a range code and an estimate in the same box in column A. Total annual on-site releases of an EPCRA Section 313 chemical from the facility of less than 1 pound may be reported in one of several ways. You should round the value to the nearest pound. If the estimate is greater than 0.5 pound, you should either enter the range code "A" for "1-10" or enter "1" in column A. If the release is equal to or less than 0.5 pounds, you may round to zero and enter "0" in column A.

Note that total annual releases of 0.5 pound or less from the processing or otherwise use of an article maintain the article status of that item. Thus, if the only releases you have are from processing an article, and such releases are equal to or less than 0.5 pound per year, you are not required to submit a report for that EPCRA Section 313 chemical. The 0.5-pound release determination does not apply to just a single article. It applies to the cumulative releases from the processing or otherwise use of the same type of article (e.g., sheet metal or plastic film) that occurs over the course of the reporting year.

Releases of 1,000 Pounds or More. For releases to any medium that amount to 1,000 pounds or more for the year, you must provide an estimate in pounds per year in column A. Any estimate provided in column A need not be reported to more than two significant figures. This estimate should be in whole numbers. Do not use decimal points.

Calculating On-Site Releases. To provide the release information in column A, EPCRA Section 313(g) (2) requires a facility to use readily available data (including monitoring data) collected pursuant to other provisions of law, or, where such data are not readily available, "reasonable estimates" of the amounts involved. If available data (including monitoring data) are known to be non-representative, facilities must make reasonable estimates using the best readily available information.

Reasonable estimates of the amounts released should be made using published emission factors, material balance calculations, or engineering calculations. You may not use emission factors or calculations to estimate releases if more accurate data are available.

No additional monitoring or measurement of the quantities or concentrations of any EPCRA Section 313 chemical released into the environment, or of the frequency of such releases, beyond that required under other provisions of law or regulation or as part of routine plant operations, is required for the purpose of completing Form R.

You must estimate the quantity (in pounds) of the EPCRA Section 313 chemical or chemical category that is released annually to each environmental medium on-site. Include only the quantity of the EPCRA Section 313 chemical in this estimate. If the EPCRA Section 313 chemical present at your facility was part of a mixture or other trade name product, calculate only the releases of the EPCRA Section 313 chemical, not the other components of the mixture or other trade name product. If you are only able to estimate the releases of the mixture or other trade name product as a whole, you should assume that the release of the EPCRA Section 313 chemical is proportional to its concentration in the mixture or other trade name product. See Part 40, Section 372.30(b) of the CFR for further information on how to calculate the concentration and weight of the EPCRA Section 313 chemical in the mixture or other trade name product.

If you are reporting an EPCRA Section 313 chemical category listed in Table II of these instructions rather than a specific EPCRA Section 313 chemical, you must combine the release data for all chemicals in the EPCRA Section 313 chemical category (e.g., all listed members of certain glycol ethers or all listed members of chlorophenols) and report the aggregate amount for that EPCRA Section 313 chemical in that category separately. For example, if your facility releases 3,000 pounds per year of 2-chlorophenol, 4,000 pounds per year of 3-chlorophenol, and 4,000 pounds per year of 4-chlorophenol to air as fugitive emissions, you must report that your facility releases 11,000 pounds per year of chlorophenols to air as fugitive emissions in Part II, Section 5.1.

For aqueous ammonia solutions, releases must be reported based on 10 percent of total aqueous ammonia. Ammonia evaporating from aqueous ammonia solutions is considered to be anhydrous ammonia; therefore, 100 percent of the anhydrous ammonia should be reported if it is released to the environment.

For dissociable nitrate compounds, release estimates should be based on the weight of the nitrate only.

For metal category compounds (e.g., chromium compounds), report releases of only the parent metal. For example, a user of various inorganic chromium salts would report the total chromium released regardless of the chemical compound and exclude any contribution to mass made by the other portion of the compound.

Section 5 Column B: Basis of Estimate

For each release and otherwise managed waste estimate (Sections 5 & 6), you are required to indicate the principal method used to determine the amount of release and otherwise managed waste reported. You should enter a letter code identifying the method that applies to the largest portion of the total estimated release and otherwise managed waste quantity.

The codes are as follows:

M1   Estimate is based on continuous monitoring data or measurements for the EPCRA Section 313 chemical.

M2   Estimate is based on periodic or random monitoring data or measurements for the EPCRA Section 313 chemical.

C   Estimate is based on mass balance calculations, such as calculation of the amount of the EPCRA Section 313 chemical in streams entering and leaving process equipment.

E1   Estimate is based on published emission factors, such as those relating release quantity to through-put or equipment type (e.g., air emission factors).

E2   Estimate is based on-site specific emission factors, such as those relating release quantity to through-put or equipment type (e.g., air emission factors).

O   Estimate is based on other approaches such as engineering calculations (e.g., estimating volatilization using published mathematical formulas) or best engineering judgment. This would include applying estimated removal efficiency to a waste stream, even if the composition of the stream before treatment was fully identified through monitoring data.

For example, if 40 percent of stack emissions of the reported EPCRA Section 313 chemical were derived using source testing data, 30 percent by mass balance, and 30 percent by published chemical-specific emission factors, you should enter the code letter "M2" for periodic or random emission monitoring.

If the monitoring data, mass balance, or emission factor used to estimate the release is not specific to

the EPCRA Section 313 chemical being reported, the form should identify the estimate based on other methods of estimation (O).

If a mass balance calculation yields the flow rate of a waste, but the quantity of reported EPCRA Section 313 chemical in the waste is based on solubility data, you should report "O" because engineering calculations were used as the basis of estimate of the quantity of the EPCRA Section 313 chemical in the waste.

If the concentration of the EPCRA Section 313 chemical in the waste was measured by continuous emissions monitoring equipment and the flow rate of the waste was determined by mass balance, then the primary basis of the estimate should be "continuous emission monitoring" (M1). Even though a mass balance calculation also contributed to the estimate, "continuous emission monitoring" should be indicated because monitoring data were used to estimate the concentration of the chemical in waste.

Mass balance (C) should only be indicated if it is directly used to calculate the mass (weight) of EPCRA Section 313 chemical released. Monitoring data should be indicated as the basis of estimate only if the EPCRA Section 313 chemical concentration is measured in the waste. Monitoring data should not be indicated, for example, if the monitoring data relate to a concentration of the EPCRA Section 313 chemical in other process streams within the facility.

It is important to realize that the accuracy and proficiency of release estimation will improve over time. However, submitters are not required to use new emission factors or estimation techniques to revise previous Form R submissions.

Section 5 Column C: Percent from Stormwater

This column relates only to Section 5.3 - discharges to receiving streams or water bodies. If your facility has monitoring data on the amount of the EPCRA Section 313 chemical in stormwater runoff (including unchanneled runoff), you must include that quantity of the EPCRA Section 313 chemical in your water release in column A and indicate the percentage of the total quantity (by weight) of the EPCRA Section 313 chemical contributed by stormwater in column C (Section 5.3C).

If your facility has monitoring data on the EPCRA Section 313 chemical and an estimate of flow rate, you must use these data to determine the percent stormwater.

If you have monitored stormwater but did not detect the EPCRA Section 313 chemical, enter zero in column C. If your facility has no stormwater monitoring data for the chemical, you should enter NA in this space on the form.

If your facility does not have periodic measurements of stormwater releases of the EPCRA Section 313 chemical, but has submitted chemical-specific monitoring data in permit applications, then these data must be used to calculate the percent contribution from stormwater. One way to calculate the flow rates from stormwater runoff is the Rational Method. In this method, flow rates, Q, can be estimated by multiplying the land area of the facility, A, by the runoff coefficient, C, and then multiplying that figure by the annual rainfall intensity, I (i.e., $Q = A \times C \times I$). The rainfall intensity, I, is specific to the geographical area of the country where the facility is located, and may be obtained from most standard engineering manuals for hydrology. The flow rate, Q, will have volumetric dimensions per unit time, and will have to be converted to units of pounds per year. The runoff coefficient represents the fraction of rainfall that does not seep into the ground but runs off as stormwater. The runoff coefficient is directly related to how the land in the drainage area is used. (See table below)

Description of Land Area	Runoff Coefficient	Description of Land Area	Runoff Coefficient
Business		Brick	0.70-0.85
Downtown areas	0.70-0.95	Drives and walks	0.70-0.85
Neighborhood areas	0.50-0.70	Roofs	0.75-0.95
Industrial		Lawns: Sandy Soil	
Light areas	0.50-0.80	Flat, 2 percent	0.05-0.10
Heavy areas	0.60-0.90	Average, 2 - 7 percent	0.10-0.15
Industrial		Steep, 7 percent	0.15-0.20
Railroad yard areas	0.20-0.40	Lawns: Heavy Soil	
Unimproved areas	0.10-0.30	Flat, 2 percent	0.13-0.17
Streets		Average, 2 - 7 percent	0.18-0.22
Asphaltic	0.70-0.95	Steep, 7 percent	0.25-0.35
Concrete	0.80-0.95		

You should choose the most appropriate runoff coefficient for your site or calculate a weighted-average coefficient, which takes into account different types of land use at your facility:

Weighted-average runoff coefficient =

(Area 1 % of total)(C1) + (Area 2 % of total)(C2) + (Area 3 % of total)(C3) + ... + (Area i % of total)(Ci)
   where

C_i =   runoff coefficient for a specific land use of Area i.

Example 13: Stormwater Runoff

Your facility is located in a semi-arid region of the United States that has an annual precipitation (including snowfall) of 12 inches of rain. (Snowfall should be converted to the equivalent inches of rain; assume one foot of snow is equivalent to one inch of rain.) The total area covered by your facility is 42 acres (about 170,000 square meters or 1,829,520 square feet). The area of your facility is 50 percent unimproved area, 10 percent asphaltic streets, and 40 percent concrete pavement.

The total stormwater runoff from your facility is therefore calculated as follows:

Land Use	% Total Area	Runoff Coefficient
Unimproved area	50	0.20
Asphaltic streets	10	0.85
Concrete pavement	40	0.90

Weighted-average runoff coefficient $= [(50\%) \times (0.20)] + [(10\%) \times (0.85)] \times [(40\%) \times (0.90)] = 0.545$

(Rainfall) $\times$ (land area) $\times$ (conversion factor) $\times$ (runoff coefficient) = stormwater runoff
(1 ft/year) $\times$ (1,829,520 ft^2) $\times$ (7.48 gal/ft^3) $\times$ (0.545) = 7,458,222 gallons/year

Total stormwater runoff = 7,458,222 gallons/year

Your stormwater monitoring data shows that the average concentration of zinc in the stormwater runoff from your facility from a biocide containing a zinc compound is 1.4 milligrams per liter. The total amount of zinc discharged to surface water through the plant wastewater discharge (non-stormwater) is 250 pounds per year. The total amount of zinc discharged with stormwater is:

(7,458,222 gallons stormwater) $\times$ (3.785 liters/gallon) = 28,229,370 liters stormwater

(28,229,370 liters stormwater) $\times$ (1.4 mg zinc/liter) $\times$ 10^3 g/mg $\times$ (1/454) lb/g = 87 lb zinc.

The total amount of zinc discharged from all sources of your facility is:

250 pounds zinc from wastewater discharged
+87 pounds zinc from stormwater runoff
337 pounds zinc total water discharged

The percentage of zinc discharge through stormwater reported in section 5.3 column C on Form R is:

(87/337) $\times$ 100% = 26%

Section 6.   Transfer(s) of the Toxic Chemical in Wastes to Off-Site Locations

You must report in this section the total annual quantity of the EPCRA Section 313 chemical in wastes sent to any off-site facility for the purposes of disposal, treatment, energy recovery, or recycling. Report the total amount of the EPCRA Section 313 chemical transferred off-site after any on-site waste treatment, recycling, or removal is completed.

For all toxic chemicals (except the dioxin and dioxin-like compounds category), do not enter the values in Section 6 in gallons, tons, liters, or any measure other than pounds. You must also enter the values as whole numbers. Numbers following a decimal point are not acceptable for toxic chemicals other than those designated as PBT chemicals. For PBT chemicals, facilities should report release and other waste management quantities greater than 0.1 pound (except the dioxin and dioxin-like compounds category) provided the accuracy and the underlying data on which the estimate is based supports this level of precision.

Dioxin and dioxin-like compounds category. Facilities should report at a level of precision supported by the accuracy of the underlying data

---

and the estimation techniques on which the estimate is based. Notwithstanding the numeric precision used when determining reporting eligibility thresholds, facilities should report on Form R to the level of accuracy that their data supports, up to seven digits to the right of the decimal. EPA's reporting software and data management systems support data precision to seven digits to the right of the decimal. The smallest quantity that needs to be reported on the Form R for the dioxin and dioxin-like compounds category is 0.0001 grams (see Example 12).

NA vs. a Numeric Value (e.g., Zero). You must enter a numeric value if you transfer an EPCRA Section 313 chemical to a Publicly Owned Treatment Works (POTW) or transfer wastes containing that toxic chemical to other off-site locations. If the aggregate amount transferred was less than 0.5 pound, then you should enter zero (unless the chemical is listed as a PBT chemical). Also report zero for transfers of listed mineral acids (i.e., hydrogen fluoride and nitric acid) if they have been neutralized to a pH of 6 or above prior to discharge to a POTW; do not check NA.

However, if you do not discharge wastewater containing the reported EPCRA Section 313 chemical to a POTW, you should check the "NA" box for the POTW's name in Section 6.1._ If you do not ship or transfer wastes containing the reported EPCRA Section 313 chemical to other off-site locations, you should check the "NA" box in Section 6.2._

Important: Number the boxes for reporting the information for each sequential POTW or other off-site location in Sections 6.1 and 6.2. In the upper left hand corner of each box, the section number is either 6.1.[ ]._.or 6.2.[ ]. This section is required only for paper filers; TRI-MEweb does this task automatically for the reporting facility.

If you report a transfer of the listed EPCRA Section 313 chemical to one or more off-site locations, POTWs, you should number the boxes in Section 6.1 as 6.1.1, 6.1.2, etc. If you transfer the EPCRA Section 313 chemical to more than one POTW, you should photocopy Page 3 of Form R as many times as necessary and then number the boxes consecutively for each POTW (e.g., 6.1.2, 6.1.3, etc.). At the bottom of each page 3 that is submitted, indicate the total number of pages numbered "3" that you are submitting as part of Form R, as well as

indicating the sequence of those pages. For example, your facility transfers the reported EPCRA Section 313 chemical in wastewaters to two POTWs. You would photocopy Page 3 once, indicate at the bottom of each Page 3 that there are a total of two pages numbered "3" and then indicate the first and second Page 3. The box for the first POTW on the first Page 3 should be numbered 6.1.1 and while the box for second POTW on the second Page 3 should be numbered 6.1.2.

If you report a transfer of the EPCRA Section 313 chemical to one or more other off-site locations, you should number the boxes in section 6.2 as 6.2.1, 6.2.2, etc. If you transfer the EPCRA Section 313 chemical to more than two other off-site locations, you should photocopy Page 4 of Form R as many times as necessary and then number the boxes consecutively for each off-site location. At the bottom of Page 4 you will find instructions for indicating the total number of Page 4s that you are submitting as part of the Form R as well as indicating the sequence of those pages. For example, your facility transfers the reported EPCRA Section 313 chemical to three other off-site locations. You should photocopy page 4 once, indicate at the bottom of Section 6.2 on each Page 4 that there are a total of two Page 4s and then indicate the first and second Page 4. The boxes for the two off-site locations on the first Page 4 would be numbered 6.2.1 and 6.2.2, while the box for the third off-site location on the second Page 4 should be numbered 6.2.3. Please note that section 6.2 starts on Page 3 and continues on Page 4.

6.1   Discharges to Publicly Owned Treatment Works

In Section 6.1.[ ], you should enter the name and address for each POTW to which your facility discharges or otherwise transfers wastewater containing the reported EPCRA Section 313 chemical. The most common transfers of this type will be conveyances of the toxic chemical in facility wastewater through underground sewage pipes; however, materials may also be trucked or transferred via some other direct methods to a POTW. In Section 6.1.[ ]A or Section 6.1.[ ]B (for columns A and B, respectively, estimate the quantity of the reported EPCRA Section 313 chemical transferred to each POTW and the basis upon which the estimate was made, respectively.

If you do not discharge wastewater containing the reported EPCRA Section 313 chemical to a POTW, enter NA in the box in Section 6.1.. (See discussion of NA vs. a Numeric Value (e.g., Zero) in the introduction of Section 6).

6.1.[ ]A. Quantity Transferred to Each POTW

Enter the total amount, in pounds, of the reported EPCRA Section 313 chemical that is contained in the wastewaters transferred to each POTW. Do not enter the total poundage of the wastewaters. If the total amount transferred is less than 1,000 pounds, you may report a range by entering the appropriate range code (range reporting in section 6.1.[ ]_A. does not apply to PBT chemicals). The following reporting range codes are to be used:

Code	Reporting Range (in pounds)
A	1-10
B	11-499
C	500-999

6.1.[ ]B   Basis of Estimate

You must identify the basis for your estimate of the total quantity of the reported EPCRA Section 313 chemical in the wastewater transferred to each POTW. You should enter one of the following letter codes that applies to the method by which the largest percentage of the estimate was derived.

M1   Estimate is based on continuous monitoring data or measurements for the EPCRA Section 313 chemical.

M2   Estimate is based on periodic or random monitoring data or measurements for the EPCRA Section 313 chemical.

C   Estimate is based on mass balance calculations, such as calculation of the amount of the EPCRA Section 313 chemical in streams entering and leaving process equipment.

E1   Estimate is based on published emission factors, such as those relating release quantity to through-put or equipment type (e.g., air emission factors).

E2   Estimate is based on-site specific emission factors, such as those relating release quantity to through-put or equipment type (e.g., air emission factors).

O   Estimate is based on other approaches such as engineering calculations (e.g., estimating volatilization using published mathematical formulas) or best engineering judgment. This would include applying estimated removal efficiency to a waste stream, even if the composition of the stream before treatment was fully identified through monitoring data.

If you transfer an EPCRA Section 313 chemical to more than one POTW, you should report the basis of estimate that was used to determine the largest percentage of the EPCRA Section 313 chemical that was transferred.

6.2   Transfers to Other Off-Site Locations

In Section 6.2 enter the EPA Identification Number (defined in 40 CFR 260.10 and therefore commonly referred to as the RCRA ID Number) name, and address for each off-site location to which your facility ships or transfers wastes containing the reported EPCRA Section 313 chemical for the purposes of disposal, treatment, energy recovery, or recycling. The RCRA ID Number may be found on the Uniform Hazardous Waste Manifest, which is required by RCRA regulations. If you ship or transfer wastes containing an EPCRA Section 313 chemical and the off-site location does not have an EPA Identification Number (e.g., it does not accept RCRA hazardous wastes) enter NA in the box for the off-site location EPA Identification Number. If you ship or transfer hazardous wastes containing an EPCRA Section 313 chemical to a facility that treats, stores, or disposes RCRA hazardous wastes, make sure to include that facility's RCRA Identification Number in the box for the off-site location EPA Identification Number. This RCRA ID is shown on the RCRA manifest that must accompany the hazardous waste to the off-site facility.

If appropriate, you must report multiple activities for each off-site location. For example, if your facility sends a reported EPCRA Section 313 chemical in a single waste stream to an off-site location where some of the EPCRA Section 313 chemical is to be recycled while the remainder of the quantity transferred is to be treated, you must report both the waste treatment and recycle activities, along with the quantity associated with each activity.

If your facility transfers an EPCRA Section 313 chemical to an off-site location and that off-site location performs more than four activities on that chemical, provide the necessary information in Box 6.2.1 for the off-site facility and the first four activities. Provide the information on the remainder of the activities in Box 6.2.2 and provide again the off-site facility identification and location information.

If you do not ship or transfer wastes containing the EPCRA Section 313 chemical to other off-site locations, you should check the NA box in Section 6.2, "Transfers to Other Off-Site Locations."

If you ship or transfer the reported EPCRA Section 313 chemical in wastes to another country, you do not need to report a RCRA ID for that waste. You should indicate NA in the RCRA ID field. Enter the complete address of the non-U.S. facility in the off-site address fields, the city in the city field, the non-U.S. state or province in the county field, the postal code in the zip code field, and the Federal Information Processing Standards (FIPS) in the country field. The most commonly used FIPS country codes are listed in Table IV. To obtain a FIPS code for a country not listed, contact the TRI Information Center. There is nothing to enter in the state field.

6.2a   Column A: Total Transfers

For each off-site location, enter the total amount, in pounds (in grams for dioxin and dioxin-like compounds), of the EPCRA Section 313 chemical that is contained in the waste transferred to that location. ***Do not enter the total quantities of the waste.*** If you do not ship or transfer wastes containing the EPCRA Section 313 chemical to other off-site locations, you should enter NA (See discussion of NA vs. a Numeric Value (e.g., Zero) in the introduction of Section 6) in the box for the off-site location's EPA Identification Number (defined in 40 CFR 260.10 and therefore commonly referred to as the RCRA ID Number).

If the total amount transferred is less than 1,000 pounds, you may report a range by entering the appropriate range code (range reporting in section 6.2 does not apply to PBT chemicals). The following reporting range codes are to be used:

Code	Reporting Range (in pounds)
A	1-10
B	11-499
C	500-999

Summary of Residue Quantities From Pilot-Scale Experimental Study[a,b]
(weight percent of drum capacity)

Unloading Method	Vessel Type	Value	Material			
			Kerosene[c]	Water[d]	Motor Oil[e]	Surfactant Solution[f]
Pumping	Steel drum	Range Mean	1.93 - 3.08 2.48	1.84 - 2.61 2.29	1.97 - 2.23 2.06	3.06 3.06
Pumping	Plastic drum	Range Mean	1.69 - 4.08 2.61	2.54 - 4.67 3.28	1.70 - 3.48 2.30	Not Available
Pouring	Bung-top steel drum	Range Mean	0.244 - 0.472 0.404	0.266 - 0.458 0.403	0.677 - 0.787 0.737	0.485 0.485
Pouring	Open-top steel drum	Range Mean	0.032 - 0.080 0.054	0.026 - 0.039 0.034	0.328 - 0.368 0.350	0.089 0.089
Gravity Drain	Slope-bottom steel tank	Range Mean	0.020 - 0.039 0.033	0.016 - 0.024 0.019	0.100 - 0.121 0.111	0.048 0.048
Gravity Drain	Dish-bottom steel tank	Range Mean	0.031 - 0.042 0.038	0.033 - 0.034 0.034	0.133 - 0.191 0.161	0.058 0.058
Gravity Drain	Dish-bottom glass-lined tank	Range Mean	0.024 - 0.049 0.040	0.020 - 0.040 0.033	0.112 - 0.134 0.127	0.040 0.040

[a] From "Releases During Cleaning of Equipment." Prepared by PEI Associates, Inc., for the U.S. Environmental Protection Agency, Office of Pesticides and Toxic Substances, Washington DC, Contract No. 68-02-4248. June 30, 1986.
[b] The values listed in this table should only be applied to similar vessel types, unloading methods, and bulk fluid materials. At viscosities greater than 200 centipoise, the residue quantities can rise dramatically and the information on this table is not applicable.
[c] For kerosene, viscosity = 5 centipoise, surface tension = 29.3 dynes/cm^2
[d] For water, viscosity = 4 centipoise, surface tension = 77.3 dynes/cm^2
[e] For motor oil, viscosity = 94 centipoise, surface tension = 34.5 dynes/cm^2
[f] For surfactant solution, viscosity = 3 centipoise, surface tension = 31.4 dynes/cm^2

Example 14: Container Residue

You have determined that a Form R for an EPCRA Section 313 chemical must be submitted. The facility purchases and uses one thousand 55-gallon steel drums that contain a 10 percent solution of the chemical. Further, it is assumed that the physical properties of the solution are similar to water. The solution is pumped from the drums directly into a mixing vessel and the "empty" drums are triple-rinsed with water. The rinse water is indirectly discharged to a POTW and the cleaned drums are sent to a drum reclaimer.

In this example, it can be assumed that all of the residual solution in the drums was transferred to the rinse water. Therefore, the quantity transferred to the drum reclaimer should be reported as "zero." The annual quantity of residual solution that is transferred to the rinse water can be estimated by multiplying the mean weight percent of residual solution remaining in water from pumping a steel drum (2.29 percent from the preceding table, "Summary of Residue Quantities From Pilot-Scale Experimental Study") by the total annual weight of solution in the drum (density of solution multiplied by drum volume). If the density is not known, it may be appropriate to use the density of water (8.34 pounds per gallon):

(2.29%) × (8.34 pounds/gallon) × (55 gallons/drum) × (1,000 drums) = 10,504 pounds solution

The concentration of the EPCRA Section 313 chemical in the solution is only 10%.

(10,504 pounds solution) × (10%) = 1,050 pounds

Therefore, 1,050 pounds of the chemical are transferred to the POTW.

Example 15: Reporting Metals and Metal Category Compounds that are sent Off-site

A facility manufactures a product containing elemental copper, exceeding the processing threshold for copper. Various metal fabrication operations for the process produce a wastewater stream that contains some residual copper and off-specification copper material. The wastewater is collected and sent directly to a POTW. Periodic monitoring data show that 500 pounds of copper were transferred to the POTW in the reporting year. The POTW eventually releases these chemicals to a stream. The off-specification products (containing copper) are collected and sent off-site to a RCRA Subtitle C landfill. Sampling analyses of the product combined with hazardous waste manifests were used to determine that 1,200 pounds of copper in the off-spec product were sent to the off-site landfill.

Therefore, the facility must report 500 pounds in Sections 6.1 and 8.1d, and 1200 pounds in Sections 6.2 (waste code M65 (RCRA Subtitle C Landfill) should be used) and 8.1d.

Note that for EPCRA Section 313 chemicals that are not metals or metal category compounds, the quantity sent for treatment at POTWs and to other off-site treatment locations must be reported in Section 8.7 - Quantity Treated Off-site. However, if you know that some or all of the chemical is not treated for destruction at the off-site location you must report that quantity in Section 8.1.

If you transfer the EPCRA Section 313 chemical in wastes to an off-site facility for distinct and multiple purposes, you must report those activities for each off-site location, along with the quantity of the reported EPCRA Section 313 chemical associated with each activity. For example, your facility transfers a total of 15,000 pounds of toluene to an off-site location that will use 5,000 pounds for the purposes of energy recovery, will enter 7,500 pounds into a recovery process, and will dispose of the remaining 2,500 pounds. These quantities and the associated activity codes must be reported separately in Section 6.2. (See Figure 5 for a hypothetical Section 6.2 completed for two off-site locations, one of which receives the transfer of 15,000 pounds of toluene as detailed.) If you have fewer than four total transfers in Section 6.2 Column A (see examples in Figure 5), an NA should be placed in Column A of the first unused row to indicate the termination of the sequence. If all four rows are used, there is no need to terminate the sequence. If there are more than four total transfers, re-enter the name of the off-site location, address, etc. in the next row (6.2.2) and then you should enter NA when the sequence has terminated if there are fewer than 8 (i.e. anytime there are fewer than 4 transfers listed in a Section 6.2 block, an NA should be used to terminate the sequence).

Do not double or multiple count amounts transferred off-site. For example, when a reported EPCRA Section 313 chemical is sent to an off-site facility for sequential activities, you should report the final disposition of the toxic chemical.

6.2b   Column B: Basis of Estimate
You must identify the basis for your estimates of the quantities of the reported EPCRA Section 313 chemical in waste transferred to each off-site location. Enter one of the following letter codes that

---

applies to the method by which the largest percentage of the estimate was derived.

M1   Estimate is based on continuous monitoring data or measurements for the EPCRA Section 313 chemical.

M2   Estimate is based on periodic or random monitoring data or measurements for the EPCRA Section 313 chemical.

C   Estimate is based on mass balance calculations, such as calculation of the amount of the EPCRA Section 313 chemical in streams entering and leaving process equipment.

E1   Estimate is based on published emission factors, such as those relating release quantity to through-put or equipment type (e.g., air emission factors).

E2   Estimate is based on site specific emission factors, such as those relating release quantity to through-put or equipment type (e.g., air emission factors).

O   Estimate is based on other approaches such as engineering calculations (e.g., estimating volatilization using published mathematical formulas) or best engineering judgment. This would include applying an estimated removal efficiency to a waste stream, even if the composition of the stream before treatment was fully identified through monitoring data.

6.2c   Column C: Type of Waste Management: Disposal/ Treatment/Energy Recovery/Recycling

You should enter one of the following M codes to identify the type of disposal, treatment, energy recovery, or recycling methods used by the off-site location for the reported EPCRA Section 313 chemical. You must use more than one line and code for a single location when distinct quantities of the reported EPCRA Section 313 chemical are subject to different waste management activities, including disposal, treatment, energy recovery, or recycling. You must use the code that represents the ultimate disposition of the chemical.

If the EPCRA Section 313 chemical is sent off-site for further direct reuse (e.g., an EPCRA Section 313 chemical in used solvent that will be used as lubricant at another facility) and does not undergo a waste management activity (i.e., release (including disposal), treatment, energy recovery, or recycling (recovery)) prior to that reuse, it need not be reported in section 6.2 or section 8.

Incineration vs. Energy Recovery

You must distinguish between incineration which is waste treatment, and legitimate energy recovery. For you to claim that a reported EPCRA Section 313 chemical sent off-site is used for the purposes of energy recovery and not for treatment for destruction, the EPCRA Section 313 chemical must have a significant heating value and must be combusted in an energy recovery unit such as an industrial boiler, furnace, or kiln. In a situation where the reported EPCRA Section 313 chemical is in a waste that is combusted in an energy recovery unit, but the EPCRA Section 313 chemical does not have a significant heating value, e.g., CFCs, you should use code M54, Incineration/Insignificant Fuel Value, to indicate that the EPCRA Section 313 chemical was incinerated in an energy recovery unit but did not contribute to the heating value of the waste.

Metals and Metal Category Compounds

Metals and metal category compounds will be managed in waste either by being released (including disposed of) or by being recycled. Remember that the release and other waste management information that you report for metal category compounds will be the total amount of the parent metal released or recycled and NOT the whole metal category compound. The metal has no heat value and thus cannot be combusted for energy recovery and cannot be treated because it cannot be destroyed. Thus, transfers of metals and metal category compounds for further waste management should be reported as either a transfer for recycling or a transfer for disposal. The applicable waste management codes for transfers of metals and metal category compounds for recycling are M24, metals recovery, M93, waste broker - recycling, or M26, other reuse/recovery. Applicable codes for transfers for disposal include M10, M41, M62, M64, M65, M66, M67, M73, M79, M81, M82, M90, M94, and M99. These codes are for off-site transfers for further waste management in which the waste stream may be treated but the metal contained in the waste stream is not treated and is ultimately released. For example, M41 should be used for a metal or metal category compound that is stabilized in preparation for disposal.

Applicable codes for Part II, Section 6.2, column C are:

Disposal

M10	Storage Only
M41	Solidification/Stabilization - Metals and Metal Category Compounds only
M62	Wastewater Treatment (Excluding POTW) - Metals and Metal Category Compounds only
M64	Other Landfills
M65	RCRA Subtitle C Landfills
M66	Subtitle C Surface Impoundment
M67	Other Surface Impoundments
M73	Land Treatment
M79	Other Land Disposal
M81	Underground Injection to Class I Wells
M82	Underground Injection to Class II-V Wells
M90	Other Off-Site Management
M94	Transfer to Waste Broker - Disposal

M99	Unknown

Treatment

M40	Solidification/Stabilization
M50	Incineration/Thermal Treatment
M54	Incineration/Insignificant Fuel Value
M61	Wastewater Treatment (Excluding POTW)
M69	Other Waste Treatment
M95	Transfer to Waste Broker - Waste Treatment

Energy Recovery

M56	Energy Recovery
M92	Transfer to Waste Broker - Energy Recovery

Recycling

M20	Solvents/Organics Recovery
M24	Metals Recovery
M26	Other Reuse or Recovery
M28	Acid Regeneration
M93	Transfer to Waste Broker - Recycling

SECTION 6. TRANSFER(S) OF THE TOXIC CHEMICAL IN WASTES TO OFF-SITE LOCATIONS						
6.1 DISCHARGES TO PUBLICLY OWNED TREATMENT WORKS (POTWs)				NA ☐		
6.1.___ POTW Name						
POTW Address						
City		County		State		ZIP

A. Quantity Transferred to this POTW (pounds/year*) (Enter range code**or estimate)	B. Basis of Estimate (Enter code)
	O

If additional pages of Part II, Section 6.1 are attached, indicate the total number of pages in this box ☐

and indicate the Part II, Section 6.1 page number in this box ☐ (Example: 1, 2, 3, etc.)

SECTION 6.2 TRANSFERS TO OTHER OFF-SITE LOCATIONS	NA ☑

6.2.___ Off-Site EPA Identification Number (RCRA ID No.)	COD56616246					
Off-Site Location Name:	Acme Waste Services					
Off-Site Address:	5 Market Street					
City Anywhere	County Hill	State CO	ZIP 80461	Country (non-US)		
Is this location under control of reporting facility or parent company?		☐ Yes	☑ No			

This off-site location receives a transfer of 15,000 pounds of toluene and will combust 5,000 pounds for the purposes of energy recovery, will enter 7,500 pounds into a recovery process, and will dispose of the remaining 2,500 pounds.

SECTION 6.2. TRANSFERS TO OTHER OFF-SITE LOCATION (CONTINUED)		
A. Total Transfer (pounds/year*) (Enter a range code** or estimate)	B. Basis of Estimate (Enter code)	C. Type of Waste Treatment/Disposal/ Recycling/Energy Recovery (Enter code)
1. 5,000	1. O	1. M 56
2. 7,500	2. C	2. M 20
3. 2,500	3. O	3. M 60
4. NA	4.	4. M

6.2.___ Off-Site EPA Identification Number (RCRA ID No.)	COD16772543					
Off-Site Location Name:	Combustion, Inc.					
Off-Site Address:	25 Facility Road					
City Dumfry	County Burns	State CO	ZIP 80500	Country (non-US)		
Is this location under control of reporting facility or parent company?	Yes ☐	No ☑				

A. Total Transfer (pounds/year*) (Enter a range code** or estimate)	B. Basis of Estimate (Enter code)	C. Type of Waste Treatment/Disposal/ Recycling/Energy Recovery (Enter code)
1. 12,500	1. O	1. M 54
2. NA	2.	2. M
3.	3.	3. M
4.	4.	4. M

This off-site location receives a transfer of 12,500 pounds of tetrachloroethylene (perchloroethylene) that is part of a waste that is combusted for the purposes of energy recovery in an industrial furnace. Note that the tetrachloroethylene should be reported using code M54 to indicate that it is combusted in an energy recovery unit but it does not contribute to the heating value of the waste.

Figure 5. Hypothetical Section 6.2 Completed for Two Off-Site Locations

Section 7. On-Site Waste Treatment, Energy Recovery, and Recycling Methods

You must report in this section the methods of waste treatment, energy recovery, and recycling applied to the reported EPCRA Section 313 chemical in wastes on-site. There are three separate sections for reporting such activities. Section 7A column c and Section 7A column e were deleted from Form R in 2005. Section 7A column d remained on the form until 2010. In 2011, column d was renamed column c which is addressed below.

Section 7A: On-Site Waste Treatment Methods and Efficiency

Most of the chemical-specific information required by EPCRA Section 313 that is reported on Form R is specific to the EPCRA Section 313 chemical rather than the waste stream containing the EPCRA Section 313 chemical. However, EPCRA Section 313 does require that waste treatment methods applied on-site to waste streams that contain the EPCRA Section 313 chemical be reported. This information is reportable regardless of whether the facility actively applies treatment or the treatment of the waste stream occurs passively. This information is collected in Section 7A of Form R.

In Section 7A, you must provide the following information if you treat waste streams containing the reported EPCRA Section 313 chemical on-site:

(a) The general waste stream types containing the EPCRA Section 313 chemical being reported;

(b) The waste treatment method(s) or sequence used on all waste streams containing the EPCRA Section 313 chemical; and

(c) The efficiency of each waste treatment method or waste treatment sequence in destroying or removing the EPCRA Section 313 chemical.

Use a separate line in Section 7A for each general waste stream type. Report only information about treatment of waste streams at your facility, not information about off-site waste treatment.

If you do not perform on-site treatment of waste streams containing the reported EPCRA Section 313 chemical, check the NA box at the top of Section 7A.

7A Column a: General Waste Stream

For each waste treatment method, indicate the type of waste stream containing the EPCRA Section 313 chemical that is treated. Enter the letter code that corresponds to the general waste stream type:

A   Gaseous (gases, vapors, airborne particulates)

W   Wastewater (aqueous waste)

L   Liquid waste streams (non-aqueous waste)

S   Solid waste streams (including sludges and slurries)

If a waste is a combination of water and organic liquid and the organic content is less than 50 percent, report it as a wastewater (W). Slurries and sludges containing water should be reported as solid waste if they contain appreciable amounts of dissolved solids, or solids that may settle, such that the viscosity or density of the waste is considerably different from that of process wastewater.

7A Column b: Waste Treatment Method(s) Sequence

Enter the appropriate waste treatment code from the list below for each on-site waste treatment method used on a waste stream containing the EPCRA Section 313 chemical, regardless of whether the waste treatment method actually removes the specific EPCRA Section 313 chemical being reported. Waste treatment methods must be reported for each type of waste stream being treated (i.e., gaseous waste streams, aqueous waste streams, liquid non-aqueous waste streams, and solids). Except for the air emission treatment codes, the waste treatment codes are not restricted to any medium.

Waste streams containing the EPCRA Section 313 chemical may have a single source or may be aggregates of many sources. For example, process water from several pieces of equipment at your facility may be combined prior to waste treatment. Report waste treatment methods that apply to the aggregate waste stream, as well as waste treatment methods that apply to individual waste streams. If your facility treats various wastewater streams containing the EPCRA Section 313 chemical in different ways, the different waste treatment methods must be listed separately.

If your facility has several pieces of equipment performing a similar service in a waste treatment sequence, you may combine the reporting for such equipment. It is not necessary to enter four codes to cover four scrubber units, for example, if all four are treating waste streams of similar character (e.g., sulfuric acid mist emissions), have similar influent concentrations, and have similar removal efficiencies. If, however, any of these parameters differs from one unit to the next, each scrubber should be listed separately.

If you are using the hardcopy paper form, and if your facility performs more than eight sequential waste treatment methods on a single general waste stream, continue listing the methods in the next row and renumber appropriately those waste treatment method code boxes you used to continue the sequence. For example, if the general waste stream in box 7A.1a had nine treatment methods applied to it, the ninth method would be indicated in the first method box for row 7A.2a. The numeral "1" would be crossed out, and a "9" would be inserted.

Treatment applied to any other general waste stream types would then be listed in the next empty row. In the scenario below, for instance, the second general waste stream would be reported in row 7A.3a. See Figure 6 for an example of a hypothetical section 7A.

Example 16: Calculating Releases and Other Waste Management Quantities

Your facility disposes of 14,000 pounds of lead chromate (PbCrO4.PbO) in an on-site landfill and transfers 16,000 pounds of lead selenite (PbSeO4) to an off-site land disposal facility. You would therefore be submitting three separate reports on the following: lead compounds, selenium compounds, and chromium compounds. However, the quantities you would be reporting would be the pounds of "parent" metal being released on-site or transferred off-site for further waste management. All quantities are based on mass balance calculations (See Section 5, Column B for information on Basis of Estimate and Section 6.2, Column C for waste management codes and information on transfers of EPCRA Section 313 chemicals in wastes). You would calculate releases of lead, chromium, and selenium by first determining the percentage by weight of these metals in the materials you use as follows:

Lead Chromate (PbCrO$_4$.PbO)   Molecular weight = 546.37
     Lead (2 Pb atoms)   Atomic weight = 207.2 × 2 = 414.4
     Chromium (1 Cr atom)   Atomic weight = 51.996

     Lead chromate is therefore (percent by weight):
         (414.4/546.37) = 75.85% lead and
         (51.996/546.37) = 9.52% chromium.

Lead Selenite (PbSeO$_4$)   Molecular weight = 350.17
     Lead (1 Pb atom)   Atomic weight = 207.2
     Selenium (1 Se atom)   Atomic weight = 78.96

     Lead selenite is therefore (percent by weight):
         (207.2/350.17) = 59.17% lead and
         (78.96/350.17) = 22.55% selenium.

The total pounds of lead, chromium, and selenium disposed of on or off-site from your facility are as follows:

Lead
Disposal on-site: 0.7585 × 14,000 = 10,619 pounds from lead chromate
Transfer off-site for disposal: 0.5917 × 16,000 = 9,467 pounds from lead selenite

Chromium
Disposal on-site: 0.0952 × 14,000 = 1,333 pounds from lead chromate

Selenium
Transfer off-site for disposal: 0.2255 × 16,000 = 3,608 pounds from lead selenite

SECTION 7A. ON-SITE WASTE TREATMENT METHODS AND EFFICIENCY								
☐ Not Applicable (NA) - Check here if no on-site waste treatment method is applied to any waste stream containing the toxic chemical or chemical category.								
a. General Waste Stream (Enter code)	b. Waste Treatment Method(s) Sequence (Enter 3- or 4-character code(s))							c. Waste Treatment Efficiency (Enter 2 character code)
7A.1a	7A.1b		1	H123	2	H124		7A.1c
W	3	H101	4	H129	5	H083		
	6	H082	7	H081	8	H075		
7A.2a	7A.2b		1	H077	2	NA		7A.2c
	3		4		5			E4
	6		7		8			
7A.3a	7A.3b		1	A01	2	NA		7A.3c
A	3		4		5			E5
	6		7		8			

Figure 6. Hypothetical Section 7A

Waste Treatment Codes

A01 Flare
A02 Condenser
A03 Scrubber
A04 Absorber
A05 Electrostatic Precipitator
A06 Mechanical Separation
A07 Other Air Emission Treatment
H040 Incineration--thermal destruction other than use as a fuel
H071 Chemical reduction with or without precipitation
H073 Cyanide destruction with or without precipitation
H075 Chemical oxidation
H076 Wet air oxidation
H077 Other chemical precipitation with or without pre-treatment
H081 Biological treatment with or without precipitation
H082 Adsorption
H083 Air or steam stripping
H101 Sludge treatment and/or dewatering
H103 Absorption
H111 Stabilization or chemical fixation prior to disposal
H112 Macro-encapsulation prior to disposal
H121 Neutralization
H122 Evaporation
H123 Settling or clarification
H124 Phase separation
H129 Other treatment

7A Column c: Waste Treatment Efficiency Estimate

In the space provided, enter the range code, based upon the codes listed below, indicating the percentage of the EPCRA Section 313 chemical removed from the waste stream through destruction, biological degradation, chemical conversion, or physical removal. The waste treatment efficiency (expressed as a range of percent removal) represents the percentage of the EPCRA Section 313 chemical destroyed or removed (based on amount or mass), not merely changes in volume or concentration of the EPCRA Section 313 chemical in the waste stream. The efficiency, which can reflect the overall removal from sequential treatment methods applied to the general waste stream, refers only to the percent destruction, degradation, conversion, or removal of the EPCRA Section 313 chemical from the waste stream; it does not refer to the percent conversion or removal of other constituents in the waste stream. The efficiency also does not refer to the general efficiency of the treatment method for any waste stream. For some waste treatment methods, the percent removal will represent removal by several mechanisms, as in an aeration basin, where an EPCRA Section 313 chemical may evaporate, biodegrade, or be physically removed from the sludge.

Percent removal can be calculated as follows:

$$\frac{(I - E)}{I} \times 100\%$$

where:

I = amount of the EPCRA Section 313 chemical in the influent waste stream (entering the waste treatment step or sequence) and

E = amount of the EPCRA Section 313 chemical in the effluent waste stream (exiting the waste treatment step or sequence).

Calculate the amount of the EPCRA Section 313 chemical in the influent waste stream by multiplying the concentration (by weight) of the EPCRA Section 313 chemical in the waste stream by the total amount or weight of the waste stream. In most cases, the percent removal compares the treated effluent to the influent for the particular type of waste stream. For solidification of wastewater, the waste treatment efficiency can be reported as code E1 (greater than 99.9999 percent) if no volatile EPCRA Section 313 chemicals were removed with the water or evaporated into the air. Percent removal does not apply to incineration because the waste stream, such as wastewater or liquids, may not exist in a comparable form after waste treatment and the purpose of incineration as a waste treatment is to destroy the EPCRA Section 313 chemical by converting it to carbon dioxide and water or other byproducts. In cases where the EPCRA Section 313 chemical is incinerated, the percent efficiency must be based on the amount of the EPCRA Section 313 chemical destroyed or combusted, except for metals or metal category compounds. In the cases in which a metal or metal category compound is incinerated, the efficiency is reported as code E6 (equal to or greater than 0 percent, but less than or equal to 50 percent).

Similarly, an efficiency of zero must be reported for any waste treatment method(s) that does not destroy, chemically convert or physically remove the EPCRA Section 313 chemical from the waste stream.

For metal category compounds, the calculation of the reportable concentration and waste treatment efficiency must be based on the weight of the parent metal, not on the weight of the metal compound. Metals are not destroyed, only physically removed or chemically converted from one form into another.

The waste treatment efficiency reported must represent only physical removal of the parent metal from the waste stream (except for incineration), not the percent chemical conversion of the metal compound. If a listed waste treatment method converts but does not remove a metal (e.g., chromium reduction), the method must be reported with a waste treatment efficiency of code E6 (equal to or greater than 0 percent, but less than or equal to 50 percent.

EPCRA Section 313 chemicals that are strong mineral acids neutralized to a pH of 6 or above are considered treated at 100 percent efficiency.

When calculating waste treatment efficiency, EPCRA Section 313(g)(2) requires a facility to use readily available data (including monitoring data) collected pursuant to other provisions of law, or, where such data are not readily available, "reasonable estimates" of the amounts involved.

Waste Treatment Efficiency Range Codes:

E1 = greater than 99.9999%
E2 = greater than 99.99%, but less than or equal to 99.9999%
E3 = greater than 99%, but less than or equal to 99.99%
E4 = greater than 95%, but less than or equal to 99%
E5 = greater than 50%, but less than or equal to 95%
E6 = equal to or greater than 0%, but less than or equal to 50%

Section 7B On-site Energy Recovery Processes
In Section 7B, you must indicate the on-site energy recovery methods used on the reported EPCRA Section 313 chemical.

EPA considers an EPCRA Section 313 chemical to be combusted for energy recovery if the toxic chemical has a significant heat value and is combusted in an energy recovery device. If a reported EPCRA Section 313 chemical is incinerated on-site but does not contribute energy to the process (e.g., chlorofluorocarbons), it must be considered waste treated on-site and reported in Section 7A. Metals and metal category compounds cannot be combusted for energy recovery and should NOT be reported in this section. Do not include the combustion of fuel oils, such as fuel oil

#6, in this section. Energy recovery may take place only in an industrial kiln, furnace, or boiler.

NA vs. a Numerical Value (e.g., Zero). If you do not perform on-site energy recovery for a waste stream that contains or contained the EPCRA Section 313 chemical, check the NA box at the top of Section 7B and enter NA in Section 8.2. If you perform on-site energy recovery for the waste stream that contains or contained the EPCRA Section 313 chemical, enter the appropriate code in Section 7B and enter the appropriate value in Section 8.2. If this quantity is less than or equal to 0.5 pound, round to zero (unless the chemical is a listed PBT chemical) and enter zero in 8.2. (Note: for metals and metal compounds, you should only report NA in Sections 7B and Section 8.2.)

Energy Recovery Codes

U01	Industrial Kiln
U02	Industrial Furnace
U03	Industrial Boiler

If your facility uses more than one on-site energy recovery method for the reported EPCRA Section 313 chemical, list the methods used in descending order (greatest to least) based on the amount of the EPCRA Section 313 chemical entering such methods.

Section 7C On-site Recycling Processes
In Section 7C, you must report the recycling methods used on the EPCRA Section 313 chemical.

In this section, use the codes below to report only the recycling methods in place at your facility that are applied to the EPCRA Section 313 chemical. Do not list any off-site recycling activities. (Information about off-site recycling must be reported in Part II, Section 6, "Transfers of the Toxic Chemical in Wastes to Off-site Locations.")

NA vs. a Numerical Value (e.g., Zero). If you do not perform on-site recycling for the reported EPCRA Section 313 chemical, check the NA box at the top of Section 7C and enter NA in Section 8.4. If you perform on-site recycling for the reported EPCRA Section 313 chemical, enter the appropriate code in Section 7C and enter the appropriate value in Section 8.4. If this quantity is less than or equal to 0.5 pound, round to zero (unless the chemical is a listed PBT chemical) and enter 0 in Section 8.4.

On-Site Recycling Codes

H10	Metal recovery (by retorting, smelting, or chemical or physical extraction
H20	Solvent recovery (including distillation, evaporation, fractionation or extraction)
H39	Other recovery or reclamation for reuse (including acid regeneration or other chemical reaction process)

If your facility uses more than one on-site recycling method for an EPCRA Section 313 chemical, enter the codes in the space provided in descending order (greatest to least) based on the volume of the reported EPCRA Section 313 chemical recovered by each process. If your facility uses more than ten separate methods for recycling the reported EPCRA Section 313 chemical on-site, then list the ten activities that recover the greatest amount of the EPCRA Section 313 chemical (again, in descending order).

Example 17: On-Site Waste Treatment

A process at the facility generates a wastewater stream containing an EPCRA Section 313 chemical (chemical A). A second process generates a wastewater stream containing two EPCRA Section 313 chemicals, a metal (chemical B) and a mineral acid (chemical C). Thresholds for all three chemicals have been exceeded and you are in the process of completing separate Form Rs for each chemical.

These two wastewater streams are combined and sent to an on-site wastewater treatment system before being discharged to a POTW. This system consists of an oil/water separator that removes 99 percent of chemical A; a neutralization tank in which the pH is adjusted to 7.5, thereby destroying 100 percent of the mineral acid (chemical C); and a settling tank where 95 percent of the metal (chemical B) is removed from the water (and eventually landfilled off-site).

Section 7A should be completed slightly differently when you file the Form R for each of the chemicals. The table accompanying this example shows how Section 7A should be completed for each chemical. First, on each Form R you should identify the type of waste stream in Section 7A.1a as wastewater (aqueous waste, code W). Next, on each Form R you should list the code for each of the treatment steps that is applied to the entire waste stream, regardless of whether the operation affects the chemical for which you are completing the Form R (for instance, the first four blocks of Section 7A.1b of all three Form Rs should show: H124 (phase separation), H121 (neutralization), H123 (settling or clarification), and N/A (to signify the end of the treatment system). Note that Section 7A.1b is not chemical specific. It applies to the entire waste stream being treated. Section 7A.1c applies to the efficiency of the entire system in destroying and/or removing the chemical for which you are preparing the Form R. You should enter E4 when filing for chemical A, E5 for chemical B, and E1 for chemical C.

Chemical A

7A.1a	7A.1b	1. H124	2. H121	7A.1c
W	3. H123	4. N/A	5.	E4
	6.	7.	8.	

Chemical B

7A.1a	7A.1b	1. H124	2. H121	7A.1c
W	3. H123	4. N/A	5.	E5
	6.	7.	8.	

Chemical C

7A.1a	7A.1b	1. H124	2. H121	7A.1c
W	3. H123	4. N/A	5.	E1
	6.	7.	8.	

Note that the *quantity* removed and/or destroyed is not reported in Section 7 and that the efficiency reported in Section 7A.1c refers to the amount of EPCRA Section 313 chemical destroyed *and/or removed* from the applicable waste stream. The amount actually destroyed should be reported in Section 8.6 (quantity treated on-site). For example, when completing the Form R for chemical B you should report "N/A" pounds in Section 8.6 because the metal has been removed from the wastewater stream, but not actually destroyed. The quantity of chemical B that is ultimately landfilled off-site should be reported in Sections 6.2 and 8.1c. However, when completing the Form R for chemical C, you should report the entire quantity in Section 8.6 because raising the pH to 7.5 will completely destroy the mineral acid.

Example 18: Reporting On-Site Energy Recovery

One waste stream generated by your facility contains, among other chemicals, toluene and Freon 113. Threshold quantities are exceeded for both of these EPCRA Section 313 chemicals, and you would, therefore, submit two separate Form R reports. This waste stream is sent to an on-site industrial furnace that uses the heat generated in a thermal hydrocarbon cracking process at your facility. Because toluene has a significant heat value (17,440 BTU/pound) and the energy is recovered in an industrial furnace, the code "U02" would be reported in Section 7B for the Form R submitted for toluene.

However, as Freon 113 does not contribute any value for energy recovery purposes, the combustion of Freon 113 in the industrial furnace is considered waste treatment, not energy recovery. You would report Freon 113 as entering a waste treatment step (i.e., incineration), in Section 7A, column b. In Section 7B the facility should report zero.

Section 8. Disposal or Other Releases, Source Reduction, and Recycling Activities

This section includes the data elements mandated by Section 6607 of the Pollution Prevention Act of 1990 (PPA).

In Section 8, you must provide information about source reduction activities and quantities of the EPCRA Section 313 chemicals managed as waste. For all appropriate questions, report only the quantity, in pounds, (or, for the dioxin and dioxin-like compounds category, grams) of the reported EPCRA Section 313 chemical itself. Do not include the weight of water, soil, or other waste constituents. When reporting on the metal category compounds, you should report only the amount of the metal portion of the compound as you do when estimating release and other waste management amounts.

Sections 8.1 through 8.9 must be completed for each EPCRA Section 313 chemical. Section 8.10 must be completed only if a source reduction activity was newly implemented specifically (in whole or in part) for the reported EPCRA Section 313 chemical during the reporting year. Section 8.11 allows you to submit additional optional information on source reduction, recycling, or pollution control activities implemented for the reported EPCRA Section 313 chemical at any time at your facility.

Sections 8.1 through 8.7 require reporting of quantities for the current reporting year, the prior year, and quantities anticipated in both the first year immediately following the reporting year and the second year following the reporting year (future estimates).

Do not enter the values in Section 8 in gallons, tons, liters, or any measure other than pounds (or, for the dioxin and dioxin-like compounds category, grams). For non-PBT chemicals, you must also enter the values as whole numbers; numbers following a decimal point are not acceptable for toxic chemicals other than those designated as PBT chemicals. For PBT chemicals (except the dioxin and dioxin-like compounds category), facilities should report release and other waste management quantities greater than 0.1 pound provided the accuracy and the underlying data on which the estimate is based supports this level of precision.

For the dioxin and dioxin-like compounds category, facilities should report at a level of precision supported by the accuracy of the underlying data and the estimation techniques on which the estimate is based. However, the smallest quantity that need be reported on the Form R for the dioxin and dioxin-like compounds category is 0.0001 grams (see Example 12). Notwithstanding the numeric precision used when determining reporting eligibility thresholds, facilities should report on Form R to the level of accuracy that their data supports, up to seven digits to the right of the decimal. EPA's reporting software and data management systems support data precision to seven digits to the right of the decimal.

NA vs. a Numeric Value (e.g., Zero). You should enter a numeric value in the relevant sections of Section 8 if your facility has released, treated, combusted for energy recovery or recycled any quantity of an EPCRA Section 313 chemical during the reporting year. If the aggregate quantity of that toxic chemical was equal to or less than 0.5 pound

---

for a particular waste management method, you should enter the value zero (unless the chemical is a PBT chemical) in the relevant section. In the case of PBTs (excluding dioxin) if the aggregate quantity of the toxic chemical is equal to or less than 0.1 pound for a particular waste management method, you should enter the value zero in the relevant section. For dioxin, if the aggregate quantity is equal to or less than .0001 grams for a particular waste management method, you should enter the value zero in the relevant section. For both PBTs and dioxin, the accuracy of the underlying data on which the estimate is based must support the specified level of precision in order to round to zero.

However, if there has been no on-site or off-site treatment, combustion for energy recovery, or recycling of the waste stream containing the EPCRA Section 313 chemical, then you should enter NA in the relevant section. (Note: for metals and metal category compounds, you should enter NA in Sections 8.2, 8.3, 8.6 and 8.7, as treatment and combustion for energy recovery generally are not applicable waste management methods for metals and metal compounds). For Section 8.1b, NA generally is not applicable recognizing the potential for spills, leaks, or fugitive emissions of the EPCRA Section 313 chemical. You should enter NA in Section 8.8 if there were no remedial actions, catastrophic events such as earthquakes, fires, or floods or one-time events not associated with normal or routine production processes for that toxic chemical. If there was a catastrophic event at your facility, but you were able to prevent any releases from occurring, then enter zero in Section 8.8.

Relationship to Other Laws

The reporting categories for quantities recycled, used for energy recovery, treated, and disposed of apply to completing Section 8 of Form R as well as to the rest of Form R. These categories are to be used only for TRI reporting. They are not intended for use in determining, under the Resource Conservation and Recovery Act (RCRA) Subtitle C regulations, whether a secondary material is a waste when recycled. These categories also do not apply to the information that may be submitted in the Biennial Report required under RCRA. In addition, these categories do not imply any future redefinition of RCRA terms and do not affect EPA's RCRA authority or authority under any other statute administered by EPA.

Differences in terminology and reporting requirements for EPCRA Section 313 chemicals reported on Form R and for hazardous wastes regulated under RCRA occur because EPCRA and the PPA focus on specific chemicals, while the RCRA regulations and the Biennial Report focus on waste streams that may include more than one chemical. For example, assume that a RCRA hazardous waste containing an EPCRA Section 313 chemical is recycled to recover certain constituents of that waste, but not the toxic chemical reported under EPCRA Section 313. The EPCRA Section 313 chemical simply passes through the recycling process and remains in the residual from the recycling process, which is disposed of. While the waste may be considered recycled under RCRA, for TRI purposes, the EPCRA Section 313 chemical constituent would be considered to be disposed of (as part of the residual from the recycling process).

An EPCRA Section 313 chemical or an EPCRA Section 313 chemical in a mixture that is a waste under RCRA must be reported in Sections 8.1 through 8.8.

Years for Which Quantities for Sections 8.1 – 8.7 Must Be Reported

Column A: Prior Year. Quantities for Sections 8.1 through 8.7 must be reported for the year immediately preceding the reporting year in column A. For reports due July 1, 2012 (reporting year 2011), the prior year is 2010. Information available at the facility that may be used to estimate the prior year's quantities include the prior year's Form R submission, supporting documentation, and recycling, energy recovery, treatment, or disposal operating logs or invoices. When reporting prior year estimates, facilities are not required to use quantities reported on the previous year's form if better information is available. TRI-MEweb prepopulates this column on the TRI form if the facility reported the previous year.

Column B: Current Reporting Year. Quantities for Sections 8.1 through 8.7 must be reported for the current reporting year in column B.

Columns C and D: Following Year and Second Following Year. Quantities for Sections 8.1 through 8.7 must be estimated for the following two years. EPA expects reasonable future quantity estimates using a logical basis. Information available at the facility to estimate quantities of the chemical expected during these years include (but are not limited to) planned source reduction activities, market projections, expected contracts, anticipated new product lines, company growth projections, and production capacity figures. Respondents should take into account protections available for trade secrets as provided in EPCRA Section 322 (42 USC 11042) for the chemical identity.

Example 19: Reporting Future Estimates

A pharmaceutical manufacturing facility uses an EPCRA Section 313 chemical in the manufacture of a prescription drug. During the reporting year (2011), the company received approval from the Food and Drug Administration to begin marketing their product as an over-the-counter drug beginning in 2012. This approval is publicly known and does not constitute confidential business information. As a result of this expanded market, the company estimates that sales and subsequent production of this drug will increase their use of the reported EPCRA Section 313 chemical by 30 percent per year for the two years following the reporting year. The facility treats the EPCRA Section 313 chemical on-site and the quantity treated is directly proportional to production activity. The facility thus estimates the total quantity of the reported EPCRA Section 313 chemical treated for the following year (2012) by adding 30 percent to the amount in column B (the amount for the current reporting year). The second following year (2013) figure can be calculated by adding an additional 30 percent to the amount reported in column C (the amount for the following year (2012) projection).

Quantities Reportable in Sections 8.1 - 8.7

Section 8 of Form R uses data collected to complete Part II, Sections 5 through 7. For this reason, Section 8 should be completed last. The relationship between Sections 5, 6, and 8.8 to Sections 8.1, 8.3, 8.5, and 8.7 are provided below in equation form.

Note on Equations. Where an equation includes a value followed by a parenthetical, this means that the equation is referring only to the portion of that value described by the parenthetical. For example, "**Section 6.2 (recycling)**" refers to the portion of the value for Section 6.2 that is recycled, while

"**Section 6.2 (treatment)**" refers to the portion of the value for Section 6.2 that is treated.

Section 8.1. In Section 8.1, facilities report disposal and other releases. This includes on-site disposal and other releases reported in Section 5 and off-site disposal and other releases reported in Section 6, but excludes quantities reported in Section 5 and 6 due to remedial actions, catastrophic events, or non-production related one-time events (see the discussion on Section 8.8). Note that EPCRA Section 329(8) defines release as "any spilling, leaking, pumping, pouring, emitting, emptying, discharging, injecting, escaping, leaching, dumping, or disposing into the environment (including the abandonment of barrels, containers, and other closed receptacles)."

Metals and metal category compounds reported in 1) Section 6.2 as sent off-site for stabilization/solidification (M41) or wastewater treatment (excluding POTWs) (M62) and/or 2) Section 6.1 - discharges to POTWs, should be reported in Section 8.1. These quantities should NOT be reported in Section 8.7 because the metals are not ultimately destroyed.

Beginning in the 2003 reporting year, Section 8.1 was divided into four Subsections (8.1a, 8.1b, 8.1c and 8.1d). Please refer to the following equations that show the relationship between Sections 5, 6, 8.8, and 8.1a through 8.1d.

Sections 8.1a and 8.1b. Toxic chemicals disposed of or otherwise released on-site are reported in 8.1a or 8.1b as appropriate. Toxic chemicals sent off-site for disposal are reported in 8.1c or 8.1d.

Section 8.1a = Section 5.4.1 + Section 5.5.1A + Section 5.5.1B - Section 8.8 (on-site disposal to landfills or UIC Class I Wells)[2]

---

[2] § 8.8 includes quantities of toxic chemicals disposed of or otherwise released on-site or managed as a waste off-site due to remedial actions, catastrophic events, or one-time events not associated with the production process. In each equation, the parenthetical following "Section 8.8" indicates which portion of § 8.8 is subtracted.

Section 8.1b = Section 5.1 + Section 5.2 + Section 5.3 + Section 5.4.2 + Section 5.5.2 + Section 5.5.3A + Section 5.5.3B + Section 5.5.4 - Section 8.8 (on-site disposal or other releases, other than disposal to landfills or UIC Class I Wells)

Sections 8.1c and 8.1d. Toxic chemicals transferred off-site to POTWs or other off-site locations and then disposed of or otherwise released should be reported in 8.1c or 8.1d as appropriate. For example, quantities of a toxic chemical sent to a landfill, or sent to a POTW and subsequently sent to a landfill are reported in Section 8.1c, while quantities of a toxic chemical sent to a surface impoundment, or sent to a POTW and subsequently released to a stream, are reported in Section 8.1d. Metals and metal category compounds sent to POTWs should be reported in one of these two sections and should not be reported as treated for destruction in Section 8.7.

Section 8.1c = Section 6.1 (portion of transfer that is untreated and ultimately disposed of in landfills or UIC Class I Wells) + Section 6.2 (quantities associated with M codes M64, M65 and M81) - Section 8.8 (off-site disposal to landfills or UIC Class I Wells)[3]

Section 8.1d = Section 6.1 (portion of transfer that is untreated and ultimately disposed of or otherwise released, other than disposal to landfills or UIC Class I Wells) + Section 6.2 (quantities associated with M codes M10, M41, M62, M66, M67, M73, M79, M82, M90, M94, and M99) - Section 8.8 (off-site disposal or other releases, other than disposal to landfills or UIC Class I Wells)[3]

Some chemicals in addition to metals and metal category compounds might not be treated for destruction at a POTW. If you know that some or all of a chemical is not treated for destruction at the POTW, you should report that quantity in Section 8.1 (as indicated in the equations above) instead of Section 8.7 (which is the quantity treated off-site).

---

[3] § 8.8 includes quantities of toxic chemicals disposed of or otherwise released on-site or managed as a waste off-site due to remedial actions, catastrophic events, or one-time events not associated with the production process. In each equation, the parenthetical following "Section 8.8" indicates which portion of § 8.8 is subtracted.

Sections 8.2 and 8.3. These relate to an EPCRA Section 313 chemical or a mixture containing an EPCRA Section 313 chemical that is used for energy recovery on-site or is sent off-site for energy recovery, unless it is a commercially available fuel (e.g., fuel oil no. 6). For the purposes of reporting on Form R, reportable on-site and off-site energy recovery is the combustion of a waste stream containing an EPCRA Section 313 chemical when:

(a)   The combustion unit is integrated into an energy recovery system (i.e., industrial furnaces, industrial kilns, and boilers); and

(b)   The EPCRA Section 313 chemical is combustible and has a significant heating value (e.g., 5000 BTU)

Note: Metals and metal category compounds cannot be combusted for energy recovery. For metals and metal category compounds, you should enter NA in Sections 8.2 and 8.3.

Quantities used for energy recovery off-site that are reported in Section 8.8 are excluded from Section 8.3.

Section 8.2 is not related to Sections 5 or 6

Section 8.3 = Section 6.2 (energy recovery) − Section 8.8 (off-site energy recovery)[3]

Sections 8.4 and 8.5. These relate to an EPCRA Section 313 chemical in a waste that is recycled on-site or is sent off-site for recycling. Quantities recycled off-site that are reported in Section 8.8 are excluded from Section 8.5.

Section 8.4 is not related to Sections 5 or 6

Section 8.5 = Section 6.2 (recycling) - Section 8.8 (off-site recycling)[3]

Section 8.6 and 8.7. These relate to an EPCRA Section 313 chemical (except for most metals and metal category compounds) or a waste containing an EPCRA Section 313 chemical that is treated for destruction on-site or is sent to a POTW or other off-site location for treatment for destruction. Most metal and category compounds are not reported in this section because they cannot be destroyed (see Appendix B). Quantities treated off-site that are reported in Section 8.8 are excluded from Section 8.7.

Section 8.6 is not related to Sections 5 or 6

Section 8.7 = Section 6.1 (portion of transfer that is ultimately treated) + Section 6.2 (treatment) - Section 8.8 (off-site treatment) [3]

Some chemicals in addition to metals and metal category compounds might not be treated for destruction at a POTW. If you know that some or all of a chemical is not treated for destruction at the POTW, you should report that quantity in Section 8.1 instead of Section 8.7. Facilities should use their best readily available information to determine the final disposition of the toxic chemical sent to the POTW, and then distribute the amount reported in Section 6.1 among Sections 8.1c, 8.1d, and 8.7, as appropriate.

Example 20: Avoiding Double-Counting Quantities in Sections 8.1 through 8.7

5,000 pounds of an EPCRA Section 313 chemical enters a treatment operation. Three thousand pounds of the EPCRA Section 313 chemical exits the treatment operation and then enters a recycling operation. Five hundred pounds of the EPCRA Section 313 chemical are in residues from the recycling operation that is subsequently sent off-site to a landfill for disposal. These quantities would be reported as follows in Section 8:

Section 8.1c: 500 pounds disposed of
Section 8.4: 2,500 pounds recycled
Section 8.6: 2,000 pounds treated (5,000 that initially entered - 3,000 that subsequently entered recycling)

To report that 5,000 pounds were treated, 3,000 pounds were recycled, and that 500 pounds were sent off-site for disposal would result in over-counting the quantities of EPCRA Section 313 chemical recycled, treated, and disposed of by 3,500 pounds.

8.8 Quantity Released to the Environment as a Result of Remedial Actions, Catastrophic Events, or One-Time Events Not Associated with Production Processes

In Section 8.8, enter the total quantity of the EPCRA Section 313 chemical disposed of or released directly into the environment or sent off-site for recycling, energy recovery, treatment, or disposal during the reporting year due to any of the following events:

(1)　remedial actions;
(2)　catastrophic events such as earthquakes, fires, or floods; or
(3)　one-time events not associated with normal or routine production processes.

These quantities should not be included in Sections 8.1, 8.3, 8.5, or 8.7.

The purpose of this section is to separate quantities recycled, used for energy recovery, treated, or released (including disposals) that are associated with normal or routine production operations from those that are not. While all quantities released, recycled, combusted for energy recovery, or treated may ultimately be preventable, this section separates the quantities that are more likely to be reduced or eliminated by process oriented source reduction activities from those releases that are largely unpredictable and are less amenable to such source reduction activities. For example, spills that occur as a routine part of production operations and could be reduced or eliminated by improved handling, loading, or unloading procedures are included in the quantities reported in Section 8.1 through 8.7 as appropriate. A total loss of containment resulting from a tank rupture caused by a tornado would be included in the quantity reported in Section 8.8.

Similarly, the amount of an EPCRA Section 313 chemical cleaned up from spills resulting from normal operations during the reporting year would not be included in Section 8.8. However, the quantity of the reported EPCRA Section 313 chemical disposed of from a remedial action (e.g., RCRA corrective action) to clean up the environmental contamination resulting from past practices should be reported in Section 8.8 because they cannot currently be addressed by source reduction methods. A remedial action for purposes of Section 8.8 is a waste cleanup (including RCRA and CERCLA operations) within the facility

---

boundary. Most remedial activities involve collecting and treating contaminated material.

Also, releases caused by catastrophic events are to be incorporated into the quantity reported in Section 8.8. Such releases may be caused by natural disasters (e.g., hurricanes and earthquakes) or by large scale accidents (e.g., fires and explosions). In addition, releases due to one-time events not associated with production (e.g., terrorist bombing) are to be included in Section 8.8. These amounts are generally unanticipated and cannot be addressed by routine process oriented accident prevention techniques. By checking your documentation for calculating estimates made for Part II, Section 5, "Quantity of the Toxic Chemical Entering Each Environmental Medium On-site," you may be able to identify disposal and release amounts from the above sources. Emergency notifications under CERCLA and EPCRA as well as accident histories required under the Clean Air Act may provide useful information. You should also check facility incident reports and maintenance records to identify one time or catastrophic events.

Note: While the information reported in Section 8.8 represents only remedial, catastrophic, or one-time events not associated with production processes, Section 5 of Form R (on-site disposal and other releases to the environment) and Section 6 (off-site transfers for further waste management) must include all on-site disposal and other releases and transfers for disposal as appropriate, regardless of whether they arise from catastrophic, remedial, or routine process operations.

Avoid Double Counting in Sections 8.1 Through 8.8

Do not double or multiple count quantities in Sections 8.1 through 8.8. The quantities reported in each of those sections should be mutually exclusive. Do not multiple count quantities entering sequential reportable activities during the reporting year.

Do not include in Sections 8.1 through 8.7 any quantities of the EPCRA Section 313 chemical disposed of or otherwise released into the environment or otherwise managed as waste off-site due to remedial actions, catastrophic events (such as earthquakes, fires, or floods), or unanticipated one-time events not associated with the production process (such as a drunk driver crashing his/her car into a drum storage area). These quantities should be reported in Section 8.8 only. For example, 10,000 pounds of diaminoanisole sulfate is released due to a catastrophic event and is subsequently treated off-site. The 10,000 pounds is reported in Section 8.8 but the amount subsequently treated off-site is not reported in Section 8.7.

Example 21: Quantity Released to the Environment as a Result of Remedial Actions, Catastrophic Events, or One-Time Events Not Associated with Production Processes.

A chemical manufacturer produces an EPCRA Section 313 chemical in a reactor that operates at low pressure. The reactants and the EPCRA Section 313 chemical product are piped in and out of the reactor at monitored and controlled temperatures. During normal operations, small amounts of fugitive emissions occur from the valves and flanges in the pipelines.

Due to a malfunction in the control panel (which is state-of-the-art and undergoes routine inspection and maintenance), the temperature and pressure in the reactor increase, the reactor ruptures, and the EPCRA Section 313 chemical is released. Because the malfunction could not be anticipated and, therefore, could not be reasonably addressed by specific source reduction activities, the amount released is included in Section 8.8. In this case, much of the EPCRA Section 313 chemical is released as a liquid and pools on the ground. It is estimated that 1,000 pounds of the EPCRA Section 313 chemical pooled on the ground and was subsequently collected and sent off-site for treatment. In addition, it is estimated that another 200 pounds of the EPCRA Section 313 chemical vaporized directly to the air from the rupture. The total amount reported in Section 8.8 is the 1,000 pounds that pooled on the ground (and subsequently sent off-site), plus the 200 pounds that vaporized into the air, a total of 1,200 pounds. The quantity sent off-site must also be reported in Section 6 (but not in Section 8.7) and the quantity that vaporized must be reported as a fugitive emission in Section 5 (but not in Section 8.1b).

8.9   Production Ratio or Activity Index

For Section 8.9, you must provide either a production ratio or an activity index. The production ratio or activity index allows year-to-year changes in release and other waste management quantities to be viewed within the context of production.

Production Ratio vs. Activity Index

A production ratio is a ratio of reporting year production to prior year production. Calculate a production ratio for the chemical when production levels most directly affect the quantity of the chemical managed as waste.

An activity index is also a ratio of current year to prior year values, but it is used when a variable other than production is the primary influence on the quantity of the reported EPCRA Section 313 chemical managed as waste. An Activity Index may be applicable when the EPCRA Section 313 chemical is "otherwise used" (e.g., non-incorporative activities such as extraction solvents, or metal degreasers).

What Variable is Used to Calculate This Ratio?

The production ratio or activity index must be based on the variable(s) that most directly affect(s) the quantity of the EPCRA Section 313 chemical generated as waste (i.e., recycled, used for energy recovery, treated, disposed or otherwise released). If an EPCRA Section 313 chemical is used in the production of refrigerators, for example, the production ratio would be based on the number of refrigerators produced (see Example 22).

In most cases, the production ratio or activity index must be based on a variable other than the quantity of the EPCRA Section 313 chemical manufactured, processed, or otherwise used. This is because indices based on chemical or material usage may reflect the effect of source reduction activities (e.g., process improvements) rather than changes in business activity. If the reported EPCRA Section 313 chemical is itself the end product, however, the quantities of the EPCRA Section 313 chemical(s) produced in the current and prior years do provide a good basis for the production ratio.

Reporting Tips:

- New for this reporting year, TRI-MEweb now includes a production ratio/activity index wizard to help you calculate your ratio automatically.

- The ratio or index must be reported to the nearest tenths or hundredths place (i.e., one or two digits to the right of the decimal point) for all EPCRA 313 chemicals, including PBT chemicals. A zero is not an acceptable response unless the calculated value is less than $1/200^{th}$, which can then be rounded to zero.

- If the manufacture, processing, or other use of the reported EPCRA Section 313 chemical began during the current reporting year, select NA as the production ratio or activity index.

- The ratio or index is not to be reported as a percent change between years (i.e., for a 10 percent increase, you would report the ratio 1.10, not 10% or 10).

- Some facilities may use the same EPCRA Section 313 chemical in more than one production process. In this case, a production ratio or activity index can be estimated by weighting the production ratio for each process based on the respective contribution of each process to the quantity of the reported EPCRA Section 313 chemical recycled, used for energy recovery, treated, or disposed of (see Example 25).

- It is important to realize that if your facility reports more than one reported EPCRA Section 313 chemical, the production ratio or activity index may vary for different chemicals.

- Details regarding the method used to calculate the Production Ratio/Activity Index can be included in Section 9.1, "Additional Information." This information will provide context for the production ratio and may help TRI data users better understand changes in releases or other waste management quantities. In Example 22, the facility could report, "Used the number of refrigerators painted as the production variable, because our facility otherwise uses toluene to paint refrigerators" in order to provide more information in Section 9.1.

Example 22: Determining a Production Ratio

Your facility's only use of toluene is as a paint carrier for a painting operation. You painted 12,000 refrigerators in the current reporting year and 10,000 refrigerators during the preceding year. The production ratio for toluene in this case is 1.2 (12,000/10,000) because the number of refrigerators produced is the primary factor determining the quantity of toluene to be reported in Sections 8.1 through 8.7.

A facility manufactures inorganic pigments, including titanium dioxide. Hydrochloric acid (acid aerosols) is produced as a waste byproduct during the production process. An appropriate production ratio for hydrochloric acid (acid aerosols) is the annual titanium dioxide production, not the amount of byproduct generated. If the facility produced 20,000 pounds of titanium dioxide during the reporting year and 26,000 pounds in the preceding year, the production ratio would be 0.77 (20,000/26,000).

Example 23: Determining an Activity Index

Your facility manufactures organic dyes in a batch process. Different colors of dyes are manufactured, and between color changes, all equipment must be thoroughly cleaned with solvent containing glycol ethers to reduce color carryover. During the preceding year, the facility produced 2,000 pounds of yellow dye in January, 9,000 pounds of green dye for February through September, 2,000 pounds of red dye in November, and another 2,000 pounds of yellow dye in December. This adds up to a total of 15,000 pounds and four color changeovers. During the reporting year, the facility produced 10,000 pounds of green dye during the first half of the year and 10,000 pounds of red dye in the second half. If your facility uses glycol ethers in this cleaning process only, an activity index of 0.5 (based on two color changeovers for the reporting year divided by four changeovers for the preceding year) is more appropriate than a production ratio of 1.33 (based on 20,000 pounds of dye produced in the current year divided by 15,000 pounds in the preceding year). In this case, an activity index, rather than a production ratio, better reflects the factors that influence the amount of solvent recycled, used for energy recovery, treated, or disposed of or released.

A facility that manufactures thermoplastic composite parts for aircraft uses toluene as a wipe solvent to clean molds. The solvent is stored in 55-gallon drums and is transferred to 1-gallon dispensers. The molds are cleaned on an as-needed basis that is not necessarily a function of the parts production rate. Operators cleaned 5,200 molds during the reporting year, but only cleaned 2,000 molds in the previous year. An activity index of 2.6 (5,200/2,000) represents the activities involving toluene usage in the facility. If the molds were cleaned after 1,000 parts were manufactured, a production ratio would equal the activity index and either could be used as the basis for the index.

A facility manufactures surgical instruments and cleans the metal parts with 1,1,1-trichloromethane in a vapor degreaser. The degreasing unit is operated in a batch mode and the metal parts are cleaned according to an irregular schedule. The activity index can be based upon the total time the metal parts are in the degreasing operation. If the degreasing unit operated 3,900 hours during the reporting year and 3,000 hours the prior year, the activity index is 1.3 (3,900/3,000).

Example 24: "NA" is Entered as the Production Ratio or Activity Index

Your facility began production of semiconductor chips during this reporting year. Perchloroethylene is used as a cleaning solvent for this operation and this is the only use of the EPCRA Section 313 chemical in your facility. You would enter NA in Section 8.9 because you have no basis of comparison in the prior year for the purposes of developing the activity index.

Example 25: Determining the Production Ratio Based on a Weighted Average

At many facilities, a reported EPCRA Section 313 chemical is used in more than one production process. In these cases, a production ratio or activity index can be estimated by weighting the production ratio for each process based on the respective contribution of each process to the quantity of the reported EPCRA Section 313 chemical recycled, used for energy recovery, treated, or disposed of.

Your facility paints bicycles with paint containing toluene. Sixteen thousand bicycles were produced in the reporting year and 14,500 were produced in the prior year. There were no significant design modifications that changed the total surface area to be painted for each bike. The bicycle production ratio is 1.1 (16,000/14,500). You estimate 12,500 pounds of toluene recycled, used for energy recovery, treated, disposed of or released as a result of bicycle production. Your facility also uses toluene as a solvent in a glue that is used to make components and add-on equipment for the bicycles. Thirteen thousand components were manufactured in the reporting year as compared to 15,000 during the prior year. The production ratio for the components using toluene is 0.87 (13,000/15,000). You estimate 1,000 pounds of toluene treated, recycled, used for energy recovery, disposed of or released as a result of components production. A production ratio can be calculated by weighting each of the production ratios based on the relative contribution each has to the quantities of toluene treated, recycled, used for energy recovery, disposed of or released during the reporting year (13,500 pounds). The production ratio is calculated as follows:

Production ratio = $1.1 \times (12,500/13,500) + 0.87 \times (1,000/13,500) = 1.08$

8.10 Did Your Facility Engage in Any Newly Implemented Source Reduction Activities for This Chemical During the Reporting Year?

Section 8.10 must be completed if a source reduction activity involving the reported EPCRA Section 313 chemical was newly implemented at your facility. An activity is considered newly implemented if it went into effect, in whole or in part, during this reporting year.

What Is Source Reduction?

Source reduction, as defined by the Pollution Prevention Act, means any practice that:

- Reduces the amount of any hazardous substance, pollutant, or contaminant entering any waste stream or otherwise released into the environment (including fugitive emissions) prior to recycling, energy recovery, treatment, or disposal; and

- Reduces the hazards to public health and the environment associated with the release of such substances, pollutants, or contaminants.

The term "source reduction" does not include any practice that alters the physical, chemical, or biological characteristics or the volume of a hazardous substance, pollutant, or contaminant through a process or activity that itself is not integral to and necessary for the production of a product or the providing of a service.

Source reduction activities include equipment or technology modifications, process or procedure modifications, reformulation or redesign of products, substitution of raw materials, and improvements in housekeeping, maintenance, training, or inventory control.

How Does Source Reduction Relate to the Quantities Reported in Sections 8.1-8.8?

Source reduction activities reduce the amount of the reported EPCRA Section 313 chemical disposed of or otherwise released (as reported in Section 8.1), used for energy recovery (as reported in Sections 8.2–8.3), recycled (as reported in Sections 8.4–8.5), or treated (as reported in Sections 8.6–8.7). Recycling, energy recovery, and treatment are not themselves considered source reduction activities because these practices occur *after* the chemical has entered a waste stream.

The focus of the section includes only those activities that are applied to reduce routine or reasonably anticipated releases or other quantities of the reported EPCRA Section 313 chemical managed as waste). Thus, you do not report in this section any activities taken to reduce or eliminate the quantities reported in Section 8.8.

Why Is Reporting on Source Reduction Activities Important?

The Pollution Prevention Act established the national policy "that pollution should be prevented or reduced at the source whenever feasible..." Reporting on source reduction activities provides important information for assessing progress towards this goal.

To promote pollution prevention, EPA has increased the prominence and accessibility of the pollution prevention information reported in Sections 8.10 and 8.11 of the Form R. For example, the 2011 TRI National Analysis highlighted the parent companies that reported the greatest number of source reduction activities. To learn more, visit www.epa.gov/TRI/P2.

How Do I Report Source Reduction Activities and Methods?

Instructions on how to report source reduction activities (as defined above) and the methods used to identify such activities are provided below.

TRI-MEweb

- **If Your Facility Implemented Source Reduction Activities.** If your facility implemented a new source reduction activity for the reported EPCRA Section 313 chemical during the reporting year, report the activity or activities that were implemented by selecting the most relevant activity code(s) from the drop down list in TRI-MEweb (see W-codes listed below).

  For each "Source Reduction Activity" reported, you must also enter one or more code(s) that correspond to the internal and external method(s) or information sources you used to identify the possibility for implementing a source reduction activity at your facility. If more than three methods were used to identify the source reduction activity, enter only the three

codes that contributed most to the decision to implement the activity.

For each source reduction code you enter in TRI-MEweb, a button to the right of the entry opens a text box that allows you to provide additional details on that source reduction practice. Similarly, to describe how each source reduction practice was identified, a button to the right of the entry opens a text box that allows you to enter additional information on the identification method(s) you selected. Optional additional information about source reduction provided via these text boxes is then added to the next section of the Form R (Section 8.11, Optional Pollution Prevention Information) preceded by the W- or T-code to which it relates.

- **If Your Facility Did Not Implement Source Reduction Activities.** If your facility did not implement any new source reduction activity for the reported EPCRA Section 313 chemical, check the "NA" box in Section 8.10. *TRI-MEweb* then provides a text box that you may use to provide information on any barriers your facility might be facing with regard to the implementation of source reduction activities. (This information is then added to your entry in Section 8.11; see Section 8.11 instructions for additional information on barriers to P2.)

Hard-Copy Reporting

- **If Your Facility Implemented Source Reduction Activities.** If using a paper form, source reduction activity codes must be entered in the first column of Sections 8.10.1 through 8.10.4. Next, indicate any methods to identify the reported source reduction activity using the T-codes provided below.

If you have fewer than four source reduction codes in Section 8.10, an NA should be placed in the first column of the first unused row to indicate the termination of the sequence. If all four rows are used, there is no need to terminate the sequence. If there are more than four source reduction codes, photocopy Page 5 of Form R as many times as necessary and then number the boxes consecutively for each source reduction activity. Enter NA when the sequence has terminated, unless the sequence ends at 4, 8, 12, 16, etc.

- **If Your Facility Did Not Implement Source Reduction Activities.** If your facility did not implement any new source reduction activity for the reported EPCRA Section 313 chemical, check the "NA" box in Section 8.10.

Source Reduction Activity Codes

Source reduction activity codes are listed below. In recent years many facilities have implemented green chemistry and green engineering practices to prevent pollution. In order to more closely represent these practices, EPA has developed six new source reduction codes. These codes are represented as: W15; W43; W50; W56; W57; and W84 and are provided in the list of source reductions below. Scenarios as to when these codes should be used are provided in Example 27.

Good Operating Practices
W13 Improved maintenance scheduling, record keeping, or procedures
W14 Changed production schedule to minimize equipment and feedstock changeovers
W15 Introduced in-line product quality monitoring or other process analysis system
W19 Other changes made in operating practices

Inventory Control
W21 Instituted procedures to ensure that materials do not stay in inventory beyond shelf-life
W22 Began to test outdated material — continue to use if still effective
W23 Eliminated shelf-life requirements for stable materials
W24 Instituted better labeling procedures
W25 Instituted clearinghouse to exchange materials that would otherwise be discarded
W29 Other changes made in inventory control

Spill and Leak Prevention
W31 Improved storage or stacking procedures
W32 Improved procedures for loading, unloading, and transfer operations
W33 Installed overflow alarms or automatic shut-off valves
W35 Installed vapor recovery systems
W36 Implemented inspection or monitoring program of potential spill or leak sources
W39 Other changes made in spill and leak prevention

Raw Material Modifications
W41 Increased purity of raw materials
W42 Substituted raw materials
W43 Substituted a feedstock or reagent chemical with a different chemical
W49 Other raw material modifications made

Process Modifications
W50 Optimized reaction conditions or otherwise increased efficiency of synthesis
W51 Instituted re-circulation within a process
W52 Modified equipment, layout, or piping
W53 Used a different process catalyst
W54 Instituted better controls on operating bulk containers to minimize discarding of empty containers
W55 Changed from small volume containers to bulk containers to minimize discarding of empty containers
W56 Reduced or eliminated use of an organic solvent
W57 Used biotechnology in manufacturing process
W58 Other process modifications made

Cleaning and Degreasing
W59 Modified stripping/cleaning equipment
W60 Changed to mechanical stripping/cleaning devices (from solvents or other materials)
W61 Changed to aqueous cleaners (from solvents or other materials)
W63 Modified containment procedures for cleaning units
W64 Improved draining procedures
W65 Redesigned parts racks to reduce drag out
W66 Modified or installed rinse systems
W67 Improved rinse equipment design
W68 Improved rinse equipment operation
W71 Other cleaning and degreasing modifications made

Surface Preparation and Finishing
W72 Modified spray systems or equipment
W73 Substituted coating materials used
W74 Improved application techniques
W75 Changed from spray to other system
W78 Other surface preparation and finishing modifications made

Product Modifications
W81 Changed product specifications
W82 Modified design or composition of product
W83 Modified packaging

W84 Developed a new chemical product to replace a previous chemical product

W89 Other product modifications made

Methods to Identify Source Reduction Activities

T01 Internal pollution prevention opportunity audit(s)

T02 External pollution prevention opportunity audit(s)

T03 Materials balance audits

T04 Participative team management

T05 Employee recommendation (independent of a formal company program

T06 Employee recommendation (under a formal company program

T07 State government technical assistance program

T08 Federal government technical assistance program

T09 Trade association/industry technical assistance program

T10 Vendor assistance

T11 Other

Example 26: Source Reduction

A facility assembles and paints furniture. Both the glue used to assemble the furniture and the paints contain EPCRA Section 313 chemicals. By examining the gluing process, the facility discovered that a new drum of glue is opened at the beginning of each shift, whether the old drum is empty or not. By adding a mechanism that prevents the drum from being changed before it is empty, the need for disposal of the glue is eliminated at the source. As a result, this activity is considered source reduction.

The painting process at this facility generates a solvent waste that contains an EPCRA Section 313 chemical that is collected and recovered. The recovered solvent is used to clean the painting equipment. The recycling activity does not reduce the amount of EPCRA Section 313 chemical recycled, and therefore is not considered a source reduction activity.

Example 27: Green Chemistry

Six new source reduction codes that could describe green chemistry and green engineering practices were added for Reporting Year 2012. These codes are listed below with a description of when to use each code to report a green chemistry or engineering activity.

W15   *Introduced in-line product quality monitoring or other process analysis system.* Select this code if the introduction of such a system led to a reduction in the amount of the EPCRA Section 313 chemical generated as waste.

W43   *Substituted a feedstock or reagent chemical with a different chemical.* Select this code if the EPCRA Section 313 chemical was a feedstock or reagent chemical and you replaced it (in whole or in part) with a different chemical.
- o   For raw material substitutions not at the level of the individual chemical (e.g., the substitution of natural gas for coal), select instead W42 *Substituted raw materials.*
- o   If use of a feedstock or reagent chemical was reduced or eliminated because of a change in the final product, select instead one of the codes listed under *Product Modifications.*

W50   *Optimized reaction conditions or otherwise increased efficiency of synthesis.* Select this code if the amount of the EPCRA Section 313 chemical generated as waste was reduced by increasing the overall efficiency of the synthesis.
- o   If efficiency of syntheses was improved by using of a different catalyst, select instead W53 *Used a different process catalyst.*

W56   *Reduced or eliminated use of an organic solvent.* Select this code if the EPCRA Section 313 chemical was used as a solvent in the process and the process was modified such that the EPCRA Section 313 chemical was either replaced or no longer used in as large a quantity.

W57   *Used biotechnology in manufacturing process.* Select this code if the use of biotechnology in the process reduced or eliminated the use of the TRI chemical.

W84   *Developed a new chemical product to replace previous chemical product.* Select this code if the EPCRA Section 313 chemical had been produced at the facility but was replaced it (in whole or in part) with the production of a different chemical or chemicals.

8.11   Optional Pollution Prevention Information

In Section 8.11, you have the opportunity to provide more detail about activities your facility undertook to reduce releases of the EPCRA Section 313 chemical, including source reduction, recycling, energy recovery, treatment or other pollution controls. EPA encourages you to provide detail in Section 8.11, as it offers your organization the opportunity to showcase its achievements in preventing pollution.

If you are using TRI-MEweb to submit your report, you can use the provided text boxes to describe your source reduction, recycling, or pollution control activities. If you are filing by paper, you may provide a description in the box provided on the Form R.

While EPA welcomes submissions about recycling and pollution control activities, the Agency is most interested in collecting information about innovative and effective source reduction activities, such as green chemistry or green engineering practices. In addition, the Agency wishes to encourage reporters to provide enough detailed information about their most effective source reduction activities to spur other facilities to adopt similar practices, as well as to inform the public about such activities being implemented in their communities.

To encourage submissions with additional pollution prevention information, EPA is increasing the prominence and accessibility of this information. Visit www.epa.gov/TRI/P2 to learn how to access this information and to view examples of optional pollution prevention information highlighted in EPA's annual TRI National Analysis report.

The following tips can help you provide meaningful additional information.

Be Specific:
- Which processes and products were affected?

- Which technologies and materials were used?
- Which release (to air, water land) or waste management quantities changed?
- Were there other benefits (e.g., costs, product quality?)
- Who provided the idea or assisted with implementation?
- Why did you implement this activity?

Enter useful URLs:
- For equipment manufacturers
- To other information sources related to the activity described

If you wish to provide additional information that is not related to pollution prevention or other environmentally friendly practices, use Section 9.1.

Barriers to Implementing Pollution Prevention Activities
You may also provide details on any barriers your facility faces in implementing additional source reduction, recycling or pollution control activities. If you choose to provide this information, EPA encourages you to describe specifically how one of these barrier categories applies to your facility:

1. Insufficient capital to install new source reduction equipment or implement new source reduction activities/initiatives.
2. Require technical information on pollution prevention techniques applicable to specific production processes.
3. Concern that product quality may decline as a result of source reduction.
4. Source reduction activities were implemented but were unsuccessful.
5. Specific regulatory/permit burdens
6. Pollution prevention previously implemented- additional reduction does not appear technically or economically feasible.
7. Other

EPA believes this information is valuable in giving a full picture of the source reduction activities your facility engages in and what barriers you face in the implementation of source reduction activities. EPA also believes this information may allow for an exchange between those that have knowledge of source reduction practices, such as the EPA P2 Program, and those that are seeking additional help. In addition, it will better enable EPA to identify those technological areas for which EPA can support basic research to identify alternative technologies that are less polluting.

9.1 Miscellaneous, Optional, and Additional Information for Your Form R Report

Your facility may provide additional information pertaining to any portion of your Form R submission in the box provided on the hard-copy form, or in the drop-down text box provided in TRI-MEweb. Your submissions to Section 9.1 regarding miscellaneous, additional, optional information may provide the Agency and/or the public with useful data that helps explain why your facility submitted data in one or more data elements that might appear unusual or inconsistent with previous TRI Form R submissions or with other data supplied by your facility during this reporting year. Such additional data may help EPA reduce the need for additional data quality control as well as additional TRI-related enforcement and compliance efforts. **Do not submit information you consider to be CBI or otherwise protected on your Form R.**

Other topics you may wish to address in Section 9.1 include:
- Changes in production levels
- Facility closures
- Staffing changes
- Calculation methods, e.g., emission factors
- Variable(s) used to calculate production ratio or activity index (Section 8.9)

D. Instructions for Completing Form R Schedule 1 (Dioxin and Dioxin-like Compounds)

D.1 What is the Form R Schedule 1?

The Form R Schedule 1 is a four-page form that mirrors the data elements from Form R Part II Chemical-Specific Information sections 5, 6, and 8 (current year only). The Form R Schedule 1 requires the reporting of the individual grams data for each member of the dioxin and dioxin-like compounds category present, and is submitted as an adjunct to the Form R. Beginning with reporting year 2008, facilities that file Form R reports for the dioxin and dioxin-like compounds category are required to determine if they have any of the information required by the new Form R Schedule 1. Facilities that have any of the information required by Form R Schedule 1 must submit a Form R Schedule 1 <u>in addition</u> to the Form R. Note: Beginning in RY2008, the listing order of the 17 members of the dioxin and dioxin-like compounds category changed.

D.2 Who is required to file a Form R Schedule 1?

Only facilities that file reports for the dioxin and dioxin-like compounds category may be required to file a Form R Schedule 1. Facilities that have any of the data required by Form R Schedule 1 for the individual members of the dioxin and dioxin-like compounds category must submit a Form R Schedule 1, in addition to the Form R. EPA notes that dioxin and dioxin-like compounds are not measured as a total quantity; the measurements are based on the individual compounds within the category. Emission factors for dioxin and dioxin-like compounds are also based on emission factors for the individual compounds within the category.

EPA's guidance document for dioxin and dioxin-like compounds (Emergency Planning And Community Right-To-Know Act - Section 313: Guidance for Reporting Toxic Chemicals within the Dioxin and Dioxin-like Compounds Category, EPA-745-B-00-021, December 2000) includes tables that contain the emission factors for the individual members of the dioxin and dioxin-like compounds category. Since measured data and emission factor data are based upon data for the individual members of the dioxin and dioxin-like compounds category, the information required by Form R Schedule 1 should be available to facilities that file Form R reports for the dioxin and dioxin-like compounds category.

D.3 What information is reported on the Form R Schedule 1?

The only data reported on the Form R Schedule 1 is the mass quantity information required in sections 5, 6, and 8 (current year only) of the Form R. All of the other information required in sections 5, 6, and 8 of the Form R (off-site location names, stream or water body names, etc.) would be the same so this information is not duplicated on Form R Schedule 1. For example, if a facility reported 5.3306 grams on Form R Section 5.1 for fugitive or non-point air emissions for the dioxin and dioxin-like compounds category then the facility would report on the Form R Schedule 1 the grams data for each individual member of the category that contributed to the 5.3306 gram total. The sum of the gram quantities reported for each individual member of the category should equal the total gram quantity reported for the category on Form R for each data element (see examples in Figure 7). The NA box has the same meaning on Form R Schedule 1 as it does on the Form R and should only be marked if it is marked on the Form R.

Form R Section 5 Example

SECTION 5. QUANTITY OF THE TOXIC CHEMICAL ENTERING EACH ENVIRONMENTAL MEDIUM ON-SITE				
		A. Total Release (pounds/year*) (Enter a range code** or estimate)	B. Basis of Estimate (Enter code)	C. Percent from Stormwater
5.1	Fugitive or non-point air emissions NA ☐	5.3306	M2	

Form R Schedule 1 Section 5 Example

SECTION 5. QUANTITY OF DIOXIN AND DIOXIN-LIKE COMPOUNDS ENTERING EACH ENVIRONMENTAL MEDIUM ON-SITE									
		5.1	NA		5.2	NA		5.3	Discharges to receiving streams or water bodies (Enter data for one stream or water body per box) NA ☐		
		Fugitive or non-point air emissions			Stack or point air emissions			5.3.1	5.3.2	5.3.3
D. Mass (grams) of each compound in the category (1-17)	1	0.0035								
	2	0.0059								
	3	0.0071								
	4	0.0008								
	5	0.0065								
	6	0.0923								
	7	0.5720								
	8	0.0723								
	9	0.0695								
	10	0.0399								
	11	0.3562								
	12	0.1309								
	13	0.0132								
	14	0.0815								
	15	1.4625								
	16	0.3126								
	17	2.1039								
If additional pages of Section 5.3 are attached, indicate the total number of pages in this box ☐									
and indicate the Section 5.3 page number in this box ☐ (Example: 1, 2, 3, etc.)									

The Form R Schedule 1 provides boxes for recording the gram quantities for all 17 individual members of the dioxin and dioxin-like compounds category. The boxes on the Form R Schedule 1 for each release type are divided into 17 boxes. Each of the boxes (1-17) corresponds to the individual members of the dioxin category as presented in Table I.

Figure 7. Hypothetical Form R, Section 5.1 and Form R Schedule 1, Section 5.1

Table I

Box #	CAS#	Chemical Name	Abbreviation
1	01746–01–6	2,3,7,8-Tetrachlorodibenzo- p-dioxin	2,3,7,8-TCDD
2	40321–76–4	1,2,3,7,8-Pentachlorodibenzo- p-dioxin	1,2,3,7,8-PeCDD
3	39227–28–6	1,2,3,4,7,8-Hexachlorodibenzo- p-dioxin	1,2,3,4,7,8-HxCDD
4	57653–85–7	1,2,3,6,7,8-Hexachlorodibenzo- p-dioxin	1,2,3,6,7,8-HxCDD
5	19408–74–3	1,2,3,7,8,9-Hexachlorodibenzo- p-dioxin	1,2,3,7,8,9-HxCDD
6	35822–46–9	1,2,3,4,6,7,8-Heptachlorodibenzo- p-dioxin	1,2,3,4,6,7,8-HpCDD
7	03268–87–9	1,2,3,4,6,7,8,9-Octachlorodibenzo- p-dioxin	1,2,3,4,6,7,8,9-OCDD
8	51207–31–9	2,3,7,8-Tetrachlorodibenzofuran	2,3,7,8-TCDF
9	57117–41–6	1,2,3,7,8-Pentachlorodibenzofuran	1,2,3,7,8-PeCDF
10	57117–31–4	2,3,4,7,8-Pentachlorodibenzofuran	2,3,4,7,8-PeCDF
11	70648–26–9	1,2,3,4,7,8-Hexachlorodibenzofuran	1,2,3,4,7,8-HxCDF
12	57117–44–9	1,2,3,6,7,8-Hexachlorodibenzofuran	1,2,3,6,7,8-HxCDF
13	72918–21–9	1,2,3,7,8,9-Hexachlorodibenzofuran	1,2,3,7,8,9-HxCDF
14	60851–34–5	2,3,4,6,7,8-Hexachlorodibenzofuran	2,3,4,6,7,8-HxCDF
15	67562–39–4	1,2,3,4,6,7,8-Heptachlorodibenzofuran	1,2,3,4,6,7,8-HpCDF
16	55673–89–7	1,2,3,4,7,8,9-Heptachlorodibenzofuran	1,2,3,4,7,8,9-HpCDF
17	39001–02–0	1,2,3,4,6,7,8,9-Octachlorodibenzofuran	1,2,3,4,6,7,8,9-OCDF

It is extremely important that facilities enter their grams data for the individual members of the category based on the box numbers in Table I (do not use the listing order from the 2007 or earlier versions of the reporting instructions). This information will be used to calculate toxic equivalency values using toxic equivalency factors that are specific to each member of the category. As with reporting on the Form R, facilities should report on the Form R Schedule 1 to the level of accuracy that their data supports, up to seven digits to the right of the decimal. EPA's reporting software and data management systems support data precision to seven digits to the right of the decimal. EPA strongly encourages facilities that report for the dioxin and dioxin-like compounds category to file their reports electronically to reduce the potential for errors when transferring the individual grams data from hard copy reports.

E. Facility Eligibility Determination for Alternate Threshold and for Reporting on TRI Form A Certification Statement

This section will help to determine whether you can submit the simplified Form A Certification Statement (hereafter referred to as Form A). The criteria are based on the total annual reportable amount of the listed chemical or chemical category and the amount manufactured, processed, or otherwise used. Note that, effective in Reporting Year 2008, the TRI Burden Reduction Rule has been voided by Congress. The criterion for using Form A has returned to what they were prior to Reporting Year 2006. The criteria are explained below. For more information about the final rule, see the TRI homepage at:
http://www.epa.gov/tri/lawsandregs/index.htm.

E.1 Alternate Threshold

On November 30, 1994, EPA published a final rule (59 FR 61488) that provides qualifying facilities an alternate threshold of 1 million pounds. Eligible facilities wishing to take advantage of this option may certify on a simplified two-page form referred to as Form A Certification Statement and do not have to use Form R. The "TRI Alternate Threshold for Facilities with Low Annual Reportable Amounts," provides facilities otherwise meeting EPCRA section 313 reporting thresholds the option of certifying on Form A provided that they do not exceed 500 pounds for the total annual reportable amount (defined below) for that chemical, and that their amounts manufactured or processed or otherwise used do not exceed one-million pounds. As with determining section 313 reporting thresholds, amounts manufactured, processed, or otherwise used are to be considered independently. This modification does not apply to forms being submitted on or before July 1, 1995 (covering the 1994 reporting year). If you fill out a Form A for an EPCRA section 313 chemical, do not fill out a Form R for that same chemical.

However, there is an exception to the alternate threshold rule described in the preceding paragraph. All PBT chemicals (except certain instances of reporting lead in stainless steel, brass or bronze alloys) are excluded from eligibility for the alternate threshold.

E.2 What is the Form A Certification Statement?

The Form A, which is described as the "certification statement" in 59 FR 61488, is intended as a means to reduce the compliance burden associated with *EPCRA section 313.* If a facility chooses to use Form A as a substitute for Form R for any eligible chemical, it must be submitted on an annual basis. Facilities wishing to take advantage of this burden reducing option may only submit Form A for chemicals that meet the conditions described in section E.1, Alternate Threshold, and should not submit a Form R to the TRI Data Processing Center for the same chemicals. The information submitted on the Form A includes facility identification information and the chemical or chemical category identity. The information submitted on the Form A will appear in the TRI data base in the same manner that information submitted on Form R appears. An approved Form A has been included in this Reporting Forms and Instructions document.

E.3 What Is the Annual Reportable Amount (ARA)?

For the purpose of this optional reporting modification, the annual reportable amount (ARA) is equal to the combined total quantities released at the facility (including disposed of within the facility), treated at the facility (as represented by amounts destroyed or converted by treatment processes), recovered at the facility as a result of recycling operations, combusted for the purpose of energy recovery at the facility, and amounts transferred from the facility to off-site locations for the purpose of recycling, energy recovery, treatment, and/or disposal. These quantities correspond to the sum of amounts reportable for data elements on EPA Form R (EPA Form 9350-1; Rev.10/09) as Part II column B of section 8, data elements 8.1 (quantity released), 8.2 (quantity used for energy recovery on-site), 8.3 (quantity used for energy recovery off-site), 8.4 (quantity recycled onsite), 8.5 (quantity recycled off-site), 8.6 (quantity treated on-site), and 8.7 (quantity treated off-site).

E.4 Recordkeeping

Each owner or operator who determines that they are eligible, and wishes to apply the alternate threshold to a particular chemical, must retain records substantiating this determination for a period of three years from the date of the submission of the Form A. These records must include sufficient documentation to support calculations as well as the calculations made by the facility that confirm their eligibility for each chemical for which the alternate threshold was applied.

A facility that fits within the category description, and manufactures, processes or otherwise uses no more than one million pounds of an EPCRA Section 313 chemical annually, and whose owner/operator elects to take advantage of the alternate threshold, is not considered an EPCRA Section 313 covered facility for that chemical for the purpose of submitting a Form R. This determination may provide further regulatory relief from other federal or state regulations that apply to facilities on the basis of their EPCRA Section 313 reporting status. A facility will need to reference other applicable regulations to determine if their actual requirements may be affected by this reporting modification.

E.5 Multi-establishment Facilities

For the purposes of using Form A, the facility must also make its determination based upon the entire facility's operations including all of its establishments (see 59 FR 61488 for greater detail). If the facility as a whole is able to take advantage of the alternate threshold, a single Form A is required. The eligibility to submit a Form A must be made on a whole facility determination. Thus, all of the information necessary to make the determination must be assembled to the facility level.

E.6 Trade Secrets

When making a trade secret claim on a Form A submission, EPA is requiring that a facility submit a unique Form A for each EPCRA Section 313 chemical meeting the conditions of the alternate threshold. Facilities may assert a trade secrecy claim for a chemical identity on the Form A as on the Form R. Reports submitted on a per chemical basis protect against the disclosure of trade secrets. Form As with trade secrecy claims, like Form Rs with similar claims, will be separately handled upon

receipt to protect against disclosure. Commingling trade secret chemical identities with non-trade secret chemical identities on the same submission increases the risk of disclosure.

> **Do not submit trade secret reports electronically.**

E.7 Metals and Metal Category Compounds

For metal category compounds, the amount applied toward the ARA is the amount of parent metal waste that is reported on Form R, but the thresholds apply to the amount of metal category compounds manufactured, processed, or otherwise used. For Form A certification involving both listed parent metals and associated metal compounds, the one million pound alternate threshold must be applied separately to the listed parent metal and the associated metal compound(s). Threshold determinations must be made independently for each because they are separately listed EPCRA Section 313 chemicals.

- If the threshold is exceeded for the listed parent metal but not the associated metal category compounds, then the releases of metal reported on Form R for the parent metal need not include the releases from the metal category compounds.

- If both the parent metal and the associated metal compounds exceed the alternate threshold, then the facility has the option of filing one Form R for both, using the metal category compound name and reporting total releases based on parent metal content.

- If neither the parent metal nor the associated metal compounds exceed the alternate threshold, then the facility must use a separate listing on Form A for each, since the reporting thresholds must be applied to each listed parent metal and all compounds in the associated compound category. EPA believes it is appropriate to make the distinction between filing the Form R and Form A because the Form R accounts for amounts of metal released or otherwise managed and Form A verifies that the alternate threshold for each listed chemical or chemical category has not been exceeded.

Similarly, separate listings on Form A must be submitted for all other listed chemicals even if EPA allows one listing on Form R to be filed for two or more listed chemicals (e.g., o-xylene, p-xylene and xylene (mixed isomers)). For example, if a facility processes in three separate process streams, xylene (mixed isomers), o-xylene, and p-xylene, and exceeds the conditions of the alternate threshold for each of these listed substances, the facility may combine the appropriate information on the o-xylene, p-xylene, and xylene (mixed isomers) into one Form R, but cannot combine the reports into one listing on Form A.

Facilities that process o-xylene, p-xylene, and xylene (mixed isomers) in separate process streams and do not exceed the conditions of the alternate threshold for one or more of the compounds may submit a separate Form A for each of the forms of xylene meeting the alternate threshold and report on Form R for those forms that do not. Similar to reporting on the parent metals and their associated category compounds described above, facilities that separately process all types (i.e., isomers) of xylene with individual activity levels within the conditions of the alternate threshold should file a separate Form A for each type of xylene.

Beginning with the 1998 reporting year, facilities may enter as many chemicals as are eligible on a single Form A Certification Statement.

F. Instructions for Completing TRI Form A Certification Statement

The following instructions provide information on how to enter data on a Form A. Some data entry fields of a Form A are done differently using the TRI's electronic reporting tool called TRI-MEweb than on a hard-copy paper form. TRI-MEweb also has automatic data quality tools that will assist users to enter valid data on their Form R and detect any errors on a Form A. This is why TRI-MEweb is EPA's preferred method to submit TRI data on a Form A Certification Statement.

Part I. Facility Identification Information

Section 1. Reporting Year

This is the calendar year to which the reported information applies, not the year in which you are submitting the report. Information for the reporting Year 2012 must be submitted on or before July 1, 2013.

Section 2. Trade Secret Information

2.1 Are you claiming the EPCRA Section 313 chemical identified on Page 3 a trade secret?

If facilities wish to report more than one eligible chemical on the same Form A, then they are not able to make trade secrecy claims. Any trade secrecy claims should be made on a separate form, and then the process is the same as using the Form R and as described in the following instructions.

The specific identity of the EPCRA Section 313 chemical being reported in Part II, Section 1, may be designated as a trade secret. If you are making a trade secret claim, mark "yes" and proceed to Section 2.2. Only check "yes" if you manufacture, process, or otherwise use the EPCRA Section 313 chemical whose identity is a trade secret. (See Page 3 of these instructions for specific information on trade secrecy claims.) If you checked "no," proceed to Section 3; do not answer Section 2.2.

2.2 If "yes" in 2.1, is this copy sanitized or unsanitized?

You should check "sanitized" if this copy of the report is the public version that does not contain the EPCRA Section 313 chemical identity but does contain a generic name that is structurally descriptive in its place, and you have claimed the EPCRA Section 313 chemical identity trade secret in Part I, Section 2.1. Otherwise, check "unsanitized."

Section 3. Certification

The Form A Certification Statement must be signed by a senior official with management responsibility for the person (or persons) completing the form. A senior management official must certify the accuracy and completeness of the information reported on the form by signing and dating the Form A. Each report must contain an original signature. Unlike the certification statement contained on Form R, the certification statement provided on the Alternate Threshold Form A pertains to the facility's eligibility of having met the conditions as described in 40 CFR Section 372.27. You should print or type in the space provided the name and title of the person who signs the statement. This certification statement applies to all the information supplied on the form and should be signed only after the form has been completed. *Failure to certify submissions will lead to a Notice of Non-Compliance. Please see Appendix C.2 Levels of Errors Identified in eFDPs.*

Section 4.   Facility Identification

4.1   Facility Name, Location, TRI Facility Identification Number and Tribal Country Name

Enter the full name that the facility presents to the public and its customers in doing business (e.g., the name that appears on invoices, signs, and other official business documents). Do not use a nickname for the facility (e.g., Main Street Plant) unless that is the legal name of the facility under which it does business. Also enter the street address, mailing address, city, county, three digit tribal code, if applicable, state, and zip code in the space provided. Do not use a post office box number as the street address. The street address provided must be the location where the EPCRA Section 313 chemicals are manufactured, processed, or otherwise used. If your mailing address and street address are the same, you should enter NA in the space for the mailing address. If the mailing address is outside of the US, include the FIPS country code, which may be found in Table IV.

If your facility is not in a county, put the name of your city, district (for example District of Columbia), or parish (if you are in Louisiana) in the county block of the Form R and Form A as well as in the County field of TRI-MEweb. "NA" or "None" are not acceptable entries. TRI-MEweb provides a drop-down menu for the county name, including city districts and parish names.

If your facility is located in Indian country as defined by 18 USC §1151, you must enter the three digit Bureau of Indian Affairs (BIA) tribal code in the "City/County/Tribe/State/ZIP code" field. The BIA tribal codes are listed in Table V of the RFI. Facilities using TRI-MEweb to complete their forms will be asked if they are located within tribal lands and, upon answering "yes", be taken to a look up table to determine the correct BIA code.

If your facility is not located in Indian country as defined by 18 USC §1151, you must enter only the city, county (as applicable), state and ZIP code. EPA will provide a default blank space for BIA code upon submitting the report to the data processing center. Facilities using TRI-MEweb to complete their forms will be asked if they are located within tribal lands and, upon answering "no", a default blank field will be entered for the tribal code field.

If you have submitted a Form A for previous reporting years, a TRI Facility Identification Number has been assigned to your facility. If you know your TRI Facility Identification Number, complete Section 4. If you do not know your TRI Facility Identification Number, contact the CDX Help Desk toll free at 1-888-890-1995. If your facility has moved, do not enter your TRI facility identification number, you should enter "New Facility."

The TRI Facility Identification Number is established by the first Form R submitted by a facility at a particular location. This identification number is retained by the facility even if the facility changes name, ownership, production processes, SIC or NAICS codes, etc. This identification number will stay with this location. If a new facility moves to this location it should use this TRI Facility Identification Number. Establishments of a facility that report separately should use the TRI Facility Identification Number of the facility.

Location information for a facility that has previously submitted data to EPA.

If your facility has submitted a Form A in previous reporting years, and is planning to submit another Form A, a TRI Facility Identification Number (TRIFID) has been already been assigned to your facility. If you know your TRIFID, you should complete Form A Section 4. If you do not know your TRI Facility Identification Number, you should visit the TRI Web page at http://www.epa.gov/tri for more information or contact your Regional TRI Program representative, or utilize Envirofacts on the Web to look up the address or facility name http://www.epa.gov/enviro and retrieve your assigned TRIFID.

Location information for a facility that has previously submitted data to EPA, but has changed physical location.

Hard-copy paper Form A Certification Statement: If your facility has moved, do not enter your previously assigned TRI Facility Identification Number (TRIFID), enter "New Facility". If you are filing a separate Form A for each establishment at your facility, you should use the same TRIFID for each establishment. You can also utilize Envirofacts on the Web to look up the address or facility name http://www.epa.gov/enviro if you are uncertain if a

TRIFID has been assigned to your new facility's location.

TRI-MEweb: If your facility has moved, you will need to request a new TRIFID to be assigned to your facility, add a new facility account to TRI-MEweb and chose to report as a new reporting facility. TRI-MEweb will automatically generate a new TRIFID to your facility. The TRIFID assigned to your new reporting facility should be used in all future reporting of TRI data.

Location information for a facility that has changed ownership, but has not changed physical location.

The TRIFID is established by the first Form A submitted by a facility at a particular location. Only a change in address warrants filing as a new facility; otherwise, the TRIFID is retained by the facility even if the facility changes name, ownership, production processes, NAICS codes, etc.

Hard-copy paper Form A: The TRIFID identification number will stay with the physical location, even if change of ownership occurs. If a new facility unit moves to this location it should use the TRIFID assigned to location. Establishments of a facility (facilities that report by part) that report separately should use the TRIFID assigned the primary facility's location.

TRI-MEweb: If your facility has changed ownership during the reporting year, the facility does not require a new TRIFID. TRI-MEweb can be used to change facility information due to change of ownership. TRI-MEweb also assists facilities and their establishments that report by part.

Location reporting TRI releases for the first time to EPA.

Hard-copy paper Form A Certification Statement: You should enter "New Facility" in the space for the TRI Facility Identification number if this is your first submission.

TRI-MEweb: If your facility is reporting for the first time, upon creating your CDX account and adding the TRI-MEweb application, you will be prompt to add a new facility account into TRI-MEweb. TRI-MEweb will automatically generate a new TRIFID to your facility. The TRIFID assigned

to your new reporting facility should be used in all future reporting of TRI data.

4.2   Federal Facility Designation

Executive Order 13423 directs federal facilities to comply with Right-To-Know Laws and Pollution Prevention Requirements. Please indicate in 4.2c. if the reporting facility is a federal facility or in 4.2d if the submitter is a contractor at a federal facility (GOCO). If the reporting facility is not a federal facility, you should leave this space blank. Form R allows a facility to report multiple submissions for the same chemical if the facility is composed of several distinct establishments. This data element provides the option of reporting full or partial facility information on Form R, however, this is not applicable for those facilities taking advantage of the Alternate Threshold and Form A.

4.3   Technical Contact

Enter the name and telephone number (including area code) of a technical representative whom EPA, state, or tribal officials may contact for clarification of the information reported on Form A. You should also enter an email address for this person. EPA encourages facilities to provide an email address for the Technical Contact on their TRI submissions because they will be able to receive important program updates and email alerts notifying them when their FDP has been updated and published for their review.

If the technical contact does not have an email address you should enter NA. This contact person does not have to be the same person who prepares the report or signs the Form A and does not necessarily need to be someone at the location of the reporting facility. However, this person should be familiar with the details of the report so that he or she can answer questions about the information provided.

4.4   Public Contact

Enter the name and telephone number (including area code) of a person who can respond to questions from the public about the form. You should also enter an e-mail address for this person. If you choose to designate the same person as both the Technical and the Public Contact, or you do not have a Public Contact, you may enter "Same as Section 4.3" in this space. This contact person does not have to be the same person who prepares the form or signs the Certification Statement and does not necessarily need to be someone at the location of the reporting facility.

4.5 North American Industry Classification System (NAICS) Code

Enter the appropriate six-digit North American Industry Classification System (NAICS) Code that is the primary NAICS Code for your facility in Section 4.5(a). (Use 2007 NAICS codes.) Enter any other applicable NAICS for your facility in 4.5 (b)-(f). If you do not know your NAICS code, consult the 2007 NAICS Manual (see Section B.2 of these instructions for ordering information) or check the SIC to NAICS crosswalk tables at http://www.census.gov.

The North American Industry Classification System (NAICS) is the economic classification system that replaced the 1987 SIC code system. A Federal Register notice was published on June 6, 2006 (71 FR 32464) adopting NAICS codes for TRI reporting. A subsequent Federal Register notice was published on June 9, 2008 (73 FR 32466) to incorporate 2007 OMB revisions and other corrections to the NAICS codes used for TRI Reporting.

4.6 Dun & Bradstreet Number(s)

Enter the nine digit number assigned by D&B for your facility or each establishment within your facility. These numbers code the facility for financial purposes. This number may be available from your facility's treasurer or financial officer. You can also obtain the numbers by calling 1-888-814-1435 or visiting this website: https://www.dnb.com/product/dlw/form_cc4.htm. If a facility does not subscribe to the D&B service, a number can be obtained, toll free at 1-800-234-3867 (8:00 AM to 6:00 PM, Local Time) or on the Web at http://www.dnb.com. If none of your establishments has been assigned a D&B number, you should enter NA in box (a). If only some of your establishments have been assigned Dun & Bradstreet numbers, enter those numbers in Part I, Section 4.6.

Section 5. Parent Company Information

You must provide information on your parent company. For TRI Reporting purposes, your parent company is as the highest level company, located in the United States, that directly owns at least 50 percent of the voting stock of your company. If there is no higher level U.S. company, select the "No U.S. Parent Company parent (for TRI reporting purposes)" check box. For example, the Bestchem Corporation is not owned or controlled by any other corporation but has sites throughout the country whose names begin with Bestchem. In this case, Bestchem Corporation should be listed as the parent company. Note that a facility that is a 50:50 joint venture is its own parent company. When a facility is owned by more than one company and there is no parent company for the entire facility (meaning that none of the facility owners directly owns at least 50 percent of the voting stock of the facility at issue), the facility should provide the name of the parent company of either the facility operator or the owner with the largest ownership interest in the facility. If neither the operator nor this owner has a parent company, then the NA box should be checked.

5.1 Name of Parent Company

Enter the name of the corporation or other business entity that is your ultimate US parent company. If there is no higher level U.S. company, select the "No U.S. Parent Company (for TRI reporting purposes)" check box.

5.2 Parent Company's Dun & Bradstreet Number

Enter the D&B number for your ultimate US parent company, if applicable. The number may be obtained from the treasurer or financial officer of the company or you may call 1-888-814-1435 or visit this website: https://www.dnb.com/product/dlw/form_cc4.htm. If your parent company does not have a D&B number, you should check the NA box.

Part II. Chemical Identification

Reporting on the Alternate Threshold Form A Certification Statement for metals, metal category compounds, and mixed isomers differs somewhat from Form R reporting. Please refer to Section E.7 for these guidelines.

Section 1.   Toxic Chemical Identity

(Important: DO NOT complete this section if you completed Section 2 of Part II below.)

1.1   CAS Number

Enter the Chemical Abstracts Service (CAS) registry number in Section 1.1 exactly as it appears in Table II of these instructions for the chemical being reported. CAS numbers are cross-referenced with an alphabetical list of chemical names in Table II. If you are reporting one of the EPCRA Section 313 chemical categories (e.g., chromium compounds), you should enter the applicable category code in the CAS number space. EPCRA Section 313 chemical category codes are listed below and can also be found in Table IIc and Appendix B.1.

EPCRA Section 313 Chemical Category Codes:

N010   Antimony compounds
N020   Arsenic compounds
N040   Barium compounds
N050   Beryllium compounds
N078   Cadmium compounds
N084   Chlorophenols
N090   Chromium compounds
N096   Cobalt compounds
N100   Copper compounds
N106   Cyanide compounds
N120   Diisocyanates
N150   Dioxin and dioxin-like compounds*
N171   Ethylenebisdithiocarbamic acid, salts and esters (EBDCs)
N230   Certain glycol ethers
N420   Lead compounds
N450   Manganese compounds
N458   Mercury compounds
N495   Nickel compounds
N503   Nicotine and salts
N511   Nitrate compounds (water dissociable; reportable only when in aqueous solution)
N575   Polybrominated biphenyls (PBBs)
N583   Polychlorinated alkanes (C10 to C13)
N590   Polycyclic aromatic compounds (PACs)

N725   Selenium compounds
N740   Silver compounds
N746   Strychnine and salts
N760   Thallium compounds
N770   Vanadium compounds
N874   Warfarin and salts
N982   Zinc compounds

* Facilities cannot take the alternate threshold for chemicals and chemical categories listed as PBT chemicals.

If you are making a trade secret claim, you must report the specific EPCRA Section 313 chemical identity on your unsanitized Form A and unsanitized substantiation form. Do not report the name of the EPCRA Section 313 chemical on your sanitized Form A or sanitized substantiation form. Include a generic name that is structurally descriptive in Part II, Section 1.3 of your sanitized Form A.

1.2   EPCRA Section 313 Chemical or Chemical Category Name

Enter the name of the EPCRA Section 313 chemical or chemical category exactly as it appears in Table II. If the EPCRA Section 313 chemical name is followed by a synonym in (parentheses), report the chemical by the name that directly follows the CAS number (i.e., not the synonym). If the EPCRA Section 313 chemical identity is actually a product trade name (e.g., dicofol), the *Chemical Abstracts 9th Collective Index* name is listed below it in brackets. You may report either name in this case.

Do not list the name of a chemical that does not appear in Table II, such as individual members of an EPCRA Section 313 chemical category. For example, if you use silver chloride, do not report silver chloride with its CAS number. Report this chemical as "silver compounds" with its category code N740.

If you are making a trade secret claim, you must report the specific EPCRA Section 313 chemical identity on your unsanitized Form A and unsanitized substantiation form. Do not report the name of the EPCRA Section 313 chemical on your sanitized Form A or sanitized substantiation form. Include a generic name in Part II, Section 1.3 of your sanitized Form A.

1.3   Generic Chemical Name

Complete Section 1.3 only if you are claiming the specific EPCRA Section 313 chemical identity of the EPCRA Section 313 chemical as a trade secret and have marked the trade secret block in Part I, Section 2.1 on Page 1 of Form A. Enter a generic chemical name that is descriptive of the chemical structure. You should limit the generic name to seventy characters (e.g., numbers, letters, spaces, punctuation) or less. Do not enter mixture names in Section 1.3; see Section 2 below.

In house plant codes and other substitute names that are not structurally descriptive of the EPCRA Section 313 chemical identity being withheld as a trade secret are not acceptable as a generic name. The generic name must appear on both sanitized and unsanitized Form A, and the name must be the same as that used on your substantiation forms.

Section 2.   Mixture Component Identity

Report the generic name provided to you by your supplier in this section if your supplier is claiming the chemical identity proprietary or trade secret. Do not answer "yes" in Part I, Section 2.1 on Page 1 of the form if you complete this section. You do not need to supply trade secret substantiation forms for this EPCRA Section 313 chemical because it is your supplier who is claiming the chemical identity a trade secret.

2.1   Generic Chemical Name Provided by Supplier

Enter the generic chemical name in this section only if the following three conditions apply:

1. You determine that the mixture contains an EPCRA Section 313 chemical but the only identity you have for that chemical is a generic name;
2. You know either the specific concentration of that EPCRA Section 313 chemical component or a maximum or average concentration level; and
3. You multiply the concentration level by the total annual amount of the whole mixture processed or otherwise used and determine that you meet the process or otherwise use threshold for that single, generically identified mixture component.

Index

Table I. NAICS Codes

1.1 NAICS codes that correspond to SIC codes 20 through 39:

113310	**Logging**
311	**Food Manufacturing**
3111	**Animal Food Manufacturing**
31111	**Animal Food Manufacturing**
311111	Dog and Cat Food Manufacturing
311119	Other Animal Food Manufacturing (except facilities primarily engaged in Custom Grain Grinding for Animal Feed)
3112	**Grain and Oilseed Milling**
31121	**Flour Milling and Malt Manufacturing**
311211	Flour Milling
311212	Rice Milling
311213	Malt Manufacturing
31122	**Starch and Vegetable Fats and Oils Manufacturing**
311221	Wet Corn Milling
311222	Soybean Processing
311223	Other Oilseed Processing
311225	Fats and Oils Refining and Blending
31123	**Breakfast Cereal Manuf.**
311230	Breakfast Cereal Manufacturing
3113	**Sugar and Confectionery Product Manufacturing**
31131	**Sugar Manufacturing**
311311	Sugarcane Mills
311312	Cane Sugar Refining
311313	Beet Sugar Manufacturing

31132 Chocolate and Confectionery Manufacturing from Cacao Beans

311320 Chocolate and Confectionery Manufacturing from Cacao Beans

31133 Confectionery Manufacturing from Purchased Chocolate

311330 Confectionery Manufacturing from Purchased Chocolate (except facilities primarily engaged in the retail sale of candy, nuts, popcorn and other confections not for immediate consumption made on the premises)

31134 Nonchocolate Confectionery Manufacturing

311340 Nonchocolate Confectionery Manufacturing (except facilities primarily engaged in the retail sale of candy, nuts, popcorn and other confections not for immediate consumption made on the premises)

3114 Fruit and Vegetable Preserving and Specialty Food Manufacturing

31141 Frozen Food Manufacturing

311411 Frozen Fruit, Juice, and Vegetable Manufacturing

311412 Frozen Specialty Food Manufacturing

31142 Fruit and Vegetable Canning, Pickling and Drying

311421 Fruit and Vegetable Canning

311422 Specialty Canning

311423 Dried and Dehydrated Food Manufacturing

3115 Dairy Product Manufacturing

31151 Dairy Product (except Frozen) Manufacturing

311511 Fluid Milk Manufacturing

311512 Creamery Butter Manufacturing

311513 Cheese Manufacturing

Table I. NAICS Codes

311514	Dry, Condensed, and Evaporated Dairy Product Manufacturing

31152 Ice Cream and Frozen Dessert Manufacturing

311520 Ice Cream and Frozen Dessert Manufacturing

3116 Animal Slaughtering and Processing

31161 Animal Slaughtering and Processing

311611 Animal (except Poultry) Slaughtering (except for facilities primarily engaged in Custom Slaughtering for individuals)

311612 Meat Processed from Carcasses [except for facilities primarily engaged in the cutting up and resale of purchased fresh carcasses for the trade (including boxed beef)]

311613 Rendering and Meat Byproduct Processing

311615 Poultry Processing

3117 Seafood Product Preparation and Packaging

311711 Seafood Canning

311712 Fresh and Frozen Seafood Processing

3118 Bakeries and Tortilla Manufacturing

31181 Bread and Bakery Product Manufacturing

311812 Commercial Bakeries

311813 Frozen Cakes, Pies, and Other Pastries Manufacturing

31182 Cookie, Cracker, and Pasta Manufacturing

311821 Cookie and Cracker Manufacturing

311822 Flour Mixes and Dough Manufacturing from Purchased Flour

311823 Dry Pasta Manufacturing

31183 Tortilla Manufacturing

311830 Tortilla Manufacturing

3119 Other Food Manufacturing

31191 Snack Food Manufacturing

311911 Roasted Nuts and Peanut Butter Manufacturing

311919 Other Snack Food Manufacturing

31192 Coffee and Tea Manufacturing

311920 Coffee and Tea Manufacturing

31193 Flavoring Syrup and Concentrate Manufacturing

311930 Flavoring Syrup and Concentrate Manufacturing

31194 Seasoning and Dressing Manufacturing

311941 Mayonnaise, Dressing, and Other Prepared Sauce Manufacturing

311942 Spice and Extract Manufacturing

31199 All Other Miscellaneous Food Manufacturing

311991 Perishable Prepared Food Manufacturing

311999 All Other Miscellaneous Food Manufacturing

312 Beverage and Tobacco Product Manufacturing

3121 Beverage Manufacturing

31211 Soft Drink and Ice Manufacturing

312111 Soft Drink Manufacturing

312112 Bottled Water Manufacturing (except facilities primarily engaged in bottling mineral or spring water)

312113 Ice Manufacturing

31212 Breweries

312120 Breweries

31213 Wineries

312130 Wineries

31214 Distilleries

Table I. NAICS Codes

312140	Distilleries		313311	Broadwoven Fabric Finishing Mills (except facilities primarily engaged in converting broadwoven piece goods and broadwoven textiles and facilities primarily engaged in sponging fabric for tailors and dressmakers)

3122   Tobacco Manufacturing

31221   Tobacco Stemming and Redrying

312210   Tobacco Stemming and Redrying

31222   Tobacco Product Manufacturing

312221   Cigarette Manufacturing

312229   Other Tobacco Product Manufacturing (except for facilities primarily engaged in providing Tobacco Sheeting Services)

313312   Textile and Fabric Finishing (except Broadwoven Fabric) Mills (except facilities primarily engaged in converting narrow woven textiles and narrow woven piece goods)

31332   Fabric Coating Mills

313320   Fabric Coating Mills

314   Textile Product Mills

313   Textile Mills

3131   Fiber, Yarn, and Thread Mills

31311   Fiber, Yarn, and Thread Mills

313111   Yarn Spinning Mills

313112   Yarn Texturizing, Throwing, and Twisting Mills

313113   Thread Mills

3132   Fabric Mills

31321   Broadwoven Fabric Mills

313210   Broadwoven Fabric Mills

31322   Narrow Fabric Mills and Schiffli Machine Embroidery

313221   Narrow Fabric Mills

313222   Schiffli Machine Embroidery

31323   Nonwoven Fabric Mills

313230   Nonwoven Fabric Mills

31324   Knit Fabric Mills

313241   Weft Knit Fabric Mills

313249   Other Knit Fabric and Lace Mills

3133   Textile and Fabric Finishing and Fabric Coating Mills

31331   Textile and Fabric Finishing Mills

3141   Textile Furnishing Mills

31411   Carpet and Rug Mills

314110   Carpet and Rug Mills

31412   Curtain and Linen Mills

314121   Curtain and Drapery Mills (except facilities primarily engaged in making custom drapery for retail sale)

314129   Other Household Textile Product Mills (except facilities primarily engaged in making custom drapery for retail sale)

3149   Other Textile Product Mills

31491   Textile Bag and Canvas Mills

314911   Textile Bag Mills

314912   Canvas and Related Product Mills

31499   All Other Textile Product Mills

314991   Rope, Cordage, and Twine Mills

314992   Tire Cord and Tire Fabric Mills

314999   All Other Miscellaneous Textile Product Mills (except facilities engaged in binding carpets and rugs for the trade, carpet cutting and binding, and embroidering on textile products (except apparel) for the trade)

315   Apparel Manufacturing

3151   Apparel Knitting Mills

31511   Hosiery and Sock Mills

Table I. NAICS Codes

315111	Sheer Hosiery Mills
315119	Other Hosiery and Sock Mills

31519 Other Apparel Knitting Mills

315191	Outerwear Knitting Mills
315192	Underwear and Nightwear Knitting Mills

3152 Cut and Sew Apparel Manufacturing

31521 Cut and Sew Apparel Contractors

315211	Men's and Boys' Cut and Sew Apparel Contractors
315212	Women's, Girls', and Infants' Cut and Sew Apparel Contractors

31522 Men's and Boys' Cut and Sew Apparel Manufacturing

315221	Men's and Boys' Cut and Sew Underwear and Nightwear Manufacturing
315222	Men's and Boys' Cut and Sew Suit, Coat, and Overcoat Manufacturing (except custom tailors primarily engaged in making and selling men's and boy's suits, cut and sewn from purchased fabric)
315223	Men's and Boys' Cut and Sew Shirt (except Work Shirt) Manufacturing (except custom tailors primarily engaged in making and selling men's and boy's dress shirts, cut and sewn from purchased fabric)
315224	Men's and Boys' Cut and Sew Trouser, Slack, and Jean Manufacturing
315225	Men's and Boys' Cut and Sew Work Clothing Manufacturing
315228	Men's and Boys' Cut and Sew Other Outerwear Manufacturing

31523 Women's and Girls' Cut and Sew Apparel Manufacturing

315231	Women's and Girls' Cut and Sew Lingerie, Loungewear, and Nightwear Manufacturing
315232	Women's and Girls' Cut and Sew Blouse and Shirt Manufacturing

315233	Women's and Girls' Cut and Sew Dress Manufacturing (except custom tailors primarily engaged in making and selling bridal dresses or gowns, or women's, misses' and girls' dresses cut and sewn from purchased fabric (except apparel contractors) (custom dressmakers)
315234	Women's and Girls' Cut and Sew Suit, Coat, Tailored Jacket, and Skirt Manufacturing
315239	Women's and Girls' Cut and Sew Other Outerwear Manufacturing

31529 Other Cut and Sew Apparel Manufacturing

315291	Infants' Cut and Sew Apparel Manufacturing
315292	Fur and Leather Apparel Manufacturing
315299	All Other Cut and Sew Apparel Manufacturing

3159 Apparel Accessories and Other Apparel Manufacturing

31599 Apparel Accessories and Other Apparel Manufacturing

315991	Hat, Cap, and Millinery Manufacturing
315992	Glove and Mitten Manufacturing
315993	Men's and Boys' Neckwear Manufacturing
315999	Other Apparel Accessories and Other Apparel Manufacturing

316 Leather and Allied Product Manufacturing

3161 Leather and Hide Tanning and Finishing

31611 Leather and Hide Tanning and Finishing

316110	Leather and Hide Tanning and Finishing

3162 Footwear Manufacturing

31621 Footwear Manufacturing

316211	Rubber and Plastics Footwear Manufacturing
316212	House Slipper Manufacturing
316213	Men's Footwear (except Athletic) Manufacturing

Table I. NAICS Codes

316214	Women's Footwear (except Athletic) Manufacturing
316219	Other Footwear Manufacturing

3169 Other Leather and Allied Product Manufacturing

31699 Other Leather and Allied Product Manufacturing

316991	Luggage Manufacturing
316992	Women's Handbag and Purse Manufacturing
316993	Personal Leather Good (except Women's Handbag and Purse) Manufacturing
316999	All Other Leather Good Manufacturing

321 Wood Product Manufacturing

3211 Sawmills and Wood Preservation

321113	Sawmills
321114	Wood Preservation

3212 Veneer, Plywood, and Engineered Wood Product Manufacturing

32121 Veneer, Plywood, and Engineered Wood Product Manufacturing

321211	Hardwood Veneer and Plywood Manufacturing
321212	Softwood Veneer and Plywood Manufacturing
321213	Engineered Wood Member (except Truss) Manufacturing
321214	Truss Manufacturing
321219	Reconstituted Wood Product Manufacturing

3219 Other Wood Product Manufacturing

32191 Millwork

321911	Wood Window and Door Manufacturing
321912	Cut Stock, Resawing Lumber, and Planing
321918	Other Millwork (including Flooring)

32192 Wood Container and Pallet Manufacturing

321920	Wood Container and Pallet Manufacturing

32199 All Other Wood Product Manufacturing

321991	Manufactured Home (Mobile Home) Manufacturing
321992	Prefabricated Wood Building Manufacturing
321999	All Other Miscellaneous Wood Product Manufacturing

322	**Paper Manufacturing**

3221 Pulp, Paper, and Paperboard Mills

32211 Pulp Mills

322110	Pulp Mills

32212 Paper Mills

322121	Paper (except Newsprint) Mills
322122	Newsprint Mills

32213 Paperboard Mills

322130	Paperboard Mills

3222 Converted Paper Product Manufacturing

32221 Paperboard Container Manufacturing

322211	Corrugated and Solid Fiber Box Manufacturing
322212	Folding Paperboard Box Manufacturing
322213	Setup Paperboard Box Manufacturing
322214	Fiber Can, Tube, Drum, and Similar Products Manufacturing
322215	Nonfolding Sanitary Food Container Manufacturing

32222 Paper Bag and Coated and Treated Paper Manufacturing

322221	Coated and Laminated Packaging Paper and Plastics Film Manufacturing

Table I. NAICS Codes

322222	Coated and Laminated Paper Manufacturing
322223	Plastics, Foil, and Coated Paper Bag Manufacturing
322224	Uncoated Paper and Multiwall Bag Manufacturing
322225	Laminated Aluminum Foil Manufacturing for Flexible Packaging Uses
322226	Surface-Coated Paperboard Manufacturing

32223 Stationery Product Manufacturing

322231	Die-Cut Paper and Paperboard Office Supplies Manufacturing
322232	Envelope Manufacturing
322233	Stationery, Tablet, and Related Product Manufacturing

32229 Other Converted Paper Product Manufacturing

322291	Sanitary Paper Product Manufacturing
322299	All Other Converted Paper Product Manufacturing

323 Printing and Related Support Activities

3231 Printing and Related Support Activities

32311 Printing

323110	Commercial Lithographic Printing
323113	Commercial Screen Printing
323114	Quick Printing (except facilities primarily engaged in reproducing text, drawings, plans, maps, or other copy by blueprinting, photocopying, mimeographing, or other methods of duplication other than printing or microfilming (*i.e.*, instant printing)
323115	Digital Printing
323116	Manifold Business Forms Printing
323117	Books Printing
323118	Blankbook, Looseleaf Binders, and Devices Manufacturing
323119	Other Commercial Printing

32312 Support Activities for Printing

323121	Tradebinding and Related Work
323122	Prepress Services

324 Petroleum and Coal Products Manufacturing

3241 Petroleum and Coal Products Manufacturing

32411 Petroleum Refineries

324110	Petroleum Refineries

32412 Asphalt Paving, Roofing, and Saturated Materials Manufacturing

324121	Asphalt Paving Mixture and Block Manufacturing
324122	Asphalt Shingle and Coating Materials Manufacturing

32419 Other Petroleum and Coal Products Manufacturing

324191	Petroleum Lubricating Oil and Grease Manufacturing
324199	All Other Petroleum and Coal Products Manufacturing

325 Chemical Manufacturing

3251 Basic Chemical Manufacturing

32511 Petrochemical Manufacturing

325110	Petrochemical Manufacturing

32512 Industrial Gas Manufacturing

325120	Industrial Gas Manufacturing

32513 Synthetic Dye and Pigment Manufacturing

325131	Inorganic Dye and Pigment Manufacturing
325132	Synthetic Organic Dye and Pigment Manufacturing

32518 Other Basic Inorganic Chemical Manufacturing

Table I. NAICS Codes

325181	Alkalies and Chlorine Manufacturing
325182	Carbon Black Manufacturing
325188	All Other Basic Inorganic Chemical Manufacturing

32519 Other Basic Organic Chemical Manufacturing

325191	Gum and Wood Chemical Manufacturing
325192	Cyclic Crude and Intermediate Manufacturing
325193	Ethyl Alcohol Manufacturing
325199	All Other Basic Organic Chemical Manufacturing

3252 Resin, Synthetic Rubber, and Artificial Synthetic Fibers and Filaments Manufacturing

32521 Resin and Synthetic Rubber Manufacturing

325211	Plastics Material and Resin Manufacturing
325212	Synthetic Rubber Manufacturing

32522 Artificial and Synthetic Fibers and Filaments Manufacturing

325221	Cellulosic Organic Fiber Manufacturing
325222	Noncellulosic Organic Fiber Manufacturing

3253 Pesticide, Fertilizer, and Other Agricultural Chemical Manufacturing

32531 Fertilizer Manufacturing

325311	Nitrogenous Fertilizer Manufacturing
325312	Phosphatic Fertilizer Manufacturing
325314	Fertilizer (Mixing Only) Manufacturing

32532 Pesticide and Other Agricultural Chemical Manufacturing

325320	Pesticide and Other Agricultural Chemical Manufacturing

3254 Pharmaceutical and Medicine Manufacturing

32541 Pharmaceutical and Medicine Manufacturing

325411	Medicinal and Botanical Manufacturing
325412	Pharmaceutical Preparation Manufacturing
325413	In-Vitro Diagnostic Substance Manufacturing
325414	Biological Product (except Diagnostic) Manufacturing

3255 Paint, Coating, and Adhesive Manufacturing

32551 Paint and Coating Manufacturing

325510	Paint and Coating Manufacturing

32552 Adhesive Manufacturing

325520	Adhesive Manufacturing

3256 Soap, Cleaning Compound, and Toilet Preparation Manufacturing

32561 Soap and Cleaning Compound Manufacturing

325611	Soap and Other Detergent Manufacturing
325612	Polish and Other Sanitation Good Manufacturing
'325613	Surface Active Agent Manufacturing

32562 Toilet Preparation Manufacturing

325620	Toilet Preparation Manufacturing

3259 Other Chemical Product and Preparation Manufacturing

32591 Printing Ink Manufacturing

325910	Printing Ink Manufacturing

32592 Explosives Manufacturing

325920	Explosives Manufacturing

32599 All Other Chemical Product and Preparation Manufacturing

Table I. NAICS Codes

325991	Custom Compounding of Purchased Resins
325992	Photographic Film, Paper, Plate, and Chemical Manufacturing
325998	All Other Miscellaneous Chemical Product and Preparation Manufacturing (except facilities primarily engaged in Aerosol can filling on a job order or contract Basis)

326 Plastics and Rubber Products Manufacturing

3261 Plastics Product Manufacturing

32611 Plastics Packaging Materials and Unlaminated Film and Sheet Manufacturing

326111	Plastics Bag Manufacturing
326112	Plastics Packaging Film and Sheet (including Laminated) Manufacturing
326113	Unlaminated Plastics Film and Sheet (except Packaging) Manufacturing

32612 Plastics, Pipe, Pipe Fitting, and Unlaminated Profile Shape Manufacturing

326121	Unlaminated Plastics Profile Shape Manufacturing
326122	Plastics Pipe and Pipe Fitting Manufacturing

32613 Laminated Plastics Plate, Sheet (except Packaging), and Shape Manufacturing

326130	Laminated Plastics Plate, Sheet (except Packaging), and Shape Manufacturing

32614 Polystyrene Foam Product Manufacturing

326140	Polystyrene Foam Product Manufacturing

32615 Urethane and Other Foam Product (except Polystyrene) Manufacturing

326150	Urethane and Other Foam Product (except Polystyrene) Manufacturing

32616 Plastics Bottle Manufacturing

326160	Plastics Bottle Manufacturing

32619 Other Plastics Product Manufacturing

326191	Plastics Plumbing Fixture Manufacturing
326192	Resilient Floor Covering Manufacturing
326199	All Other Plastics Product Manufacturing

3262 Rubber Product Manufacturing

32621 Tire Manufacturing

326211	Tire Manufacturing (except Retreading)

32622 Rubber and Plastics Hoses and Belting Manufacturing

326220	Rubber and Plastics Hoses and Belting Manufacturing

32629 Other Rubber Product Manufacturing

326291	Rubber Product Manufacturing for Mechanical Use
326299	All Other Rubber Product Manufacturing

327 Nonmetallic Mineral Product Manufacturing

3271 Clay Product and Refractory Manufacturing

32711 Pottery, Ceramics, and Plumbing Fixture Manufacturing

327111	Vitreous China Plumbing Fixture and China and Earthenware Bathroom Accessories Manufacturing
327112	Vitreous China, Fine Earthenware, and Other Pottery Product Manufacturing (except facilities primarily engaged in manufacturing and selling pottery on site)
327113	Porcelain Electrical Supply Manufacturing

Table I. NAICS Codes

32712	**Clay Building Material and Refractories Manufacturing**
327121	Brick and Structural Clay Tile Manufacturing
327122	Ceramic Wall and Floor Tile Manufacturing
327123	Other Structural Clay Product Manufacturing
327124	Clay Refractory Manufacturing
327125	Nonclay Refractory Manufacturing

3272	**Glass and Glass Product Manufacturing**

32721	**Glass and Glass Product Manufacturing**
327211	Flat Glass Manufacturing
327212	Other Pressed and Blown Glass and Glassware Manufacturing
327213	Glass Container Manufacturing
327215	Glass Product Manufacturing Made of Purchased Glass

3273	**Cement and Concrete Product Manufacturing**

32731	**Cement Manufacturing**
327310	Cement Manufacturing

32732	**Ready-Mix Concrete Manufacturing**
327320	Ready-Mix Concrete Manufacturing

32733	**Concrete, Pipe, Brick, and Block Manufacturing**
327331	Concrete Block and Brick Manufacturing
327332	Concrete Pipe Manufacturing

32739	**Other Concrete Product Manufacturing**
327390	Other Concrete Product Manufacturing

3274	**Lime and Gypsum Product Manufacturing**

32741	**Lime Manufacturing**
327410	Lime Manufacturing

32742	**Gypsum Product Manufacturing**
327420	Gypsum Product Manufacturing

3279	**Other Nonmetallic Mineral Product Manufacturing**

32791	**Abrasive Product Manufacturing**
327910	Abrasive Product Manufacturing

32799	**All Other Nonmetallic Mineral Product Manufacturing**
327991	Cut Stone and Stone Product Manufacturing
327992	Ground or Treated Mineral and Earth Manufacturing
327993	Mineral Wool Manufacturing
327999	All Other Miscellaneous Nonmetallic Mineral Product Manufacturing

331	**Primary Metal Manufacturing**

3311	**Iron and Steel Mills and Ferroalloy Manufacturing**

33111	**Iron and Steel Mills and Ferroalloy Manufacturing**
331111	Iron and Steel Mills
331112	Electrometallurgical Ferroalloy Product Manufacturing

3312	**Steel Product Manufacturing from Purchased Steel**

33121	**Iron and Steel Pipe and Tube Manufacturing from Purchased Steel**
331210	Iron and Steel Pipe and Tube Manufacturing from Purchased Steel

33122	**Rolling and Drawing of Purchased Steel**
331221	Rolled Steel Shape Manufacturing
331222	Steel Wire Drawing

Table I. NAICS Codes

3313	**Alumina and Aluminum Production and Processing**
33131	**Alumina and Aluminum Production and Processing**
331311	Alumina Refining
331312	Primary Aluminum Production
331314	Secondary Smelting and Alloying of Aluminum
331315	Aluminum Sheet, Plate, and Foil Manufacturing
331316	Aluminum Extruded Product Manufacturing
331319	Other Aluminum Rolling and Drawing
3314	**Nonferrous Metal (except Aluminum) Production and Processing**
33141	**Nonferrous Metal (except Aluminum) Smelting and Refining**
331411	Primary Smelting and Refining of Copper
331419	Primary Smelting and Refining of Nonferrous Metal (except Copper and Aluminum)
33142	**Copper Rolling, Drawing, Extruding and Alloying**
331421	Copper Rolling, Drawing, and Extruding
331422	Copper Wire (except Mechanical) Drawing
331423	Secondary Smelting, Refining, and Alloying of Copper
33149	**Nonferrous Metal (except Copper and Aluminum) Rolling, Drawing, Extruding, and Alloying**
331491	Nonferrous Metal (except Copper and Aluminum) Rolling, Drawing, and Extruding
331492	Secondary Smelting, Refining, and Alloying of Nonferrous Metal (except Copper and Aluminum)
3315	**Foundries**

33151	**Ferrous Metal Foundries**
331511	Iron Foundries
331512	Steel Investment Foundries
331513	Steel Foundries (except Investment)
33152	**Nonferrous Metal Foundries**
331521	Aluminum Die-Casting Foundries
331522	Nonferrous (except Aluminum) Die-Casting Foundries
331524	Aluminum Foundries (except Die-Casting)
331525	Copper Foundries (except Die-Casting)
331528	Other Nonferrous Foundries (except Die-Casting)
332	**Fabricated Metal Product Manufacturing**
3321	**Forging and Stamping**
33211	**Forging and Stamping**
332111	Iron and Steel Forging
332112	Nonferrous Forging
332114	Custom Roll Forming
332115	Crown and Closure Manufacturing
332116	Metal Stamping
332117	Powder Metallurgy Part Manufacturing
3322	**Cutlery and Handtool Manufacturing**
33221	**Cutlery and Handtool Manufacturing**
332211	Cutlery and Flatware (except Precious) Manufacturing
332212	Hand and Edge Tool Manufacturing
332213	Saw Blade and Handsaw Manufacturing
332214	Kitchen Utensil, Pot, and Pan Manufacturing
3323	**Architectural and Structural Metals Manufacturing**

Table I. NAICS Codes

33231 Plate Work and Fabricated Structural Product Manufacturing

332311 Prefabricated Metal Building and Component Manufacturing

332312 Fabricated Structural Metal Manufacturing

332313 Plate Work Manufacturing

33232 Ornamental and Architectural Metal Products Manufacturing

332321 Metal Window and Door Manufacturing

332322 Sheet Metal Work Manufacturing

332323 Ornamental and Architectural Metal Work Manufacturing

3324 Boiler, Tank, and Shipping Container Manufacturing

33241 Power Boiler and Heat Exchanger Manufacturing

332410 Power Boiler and Heat Exchanger Manufacturing

33242 Metal Tank (Heavy Gauge) Manufacturing

332420 Metal Tank (Heavy Gauge) Manufacturing

33243 Metal Can, Box, and Other Metal Container (Light Gauge) Manufacturing

332431 Metal Can Manufacturing

332439 Other Metal Container Manufacturing

3325 Hardware Manufacturing

33251 Hardware Manufacturing

332510 Hardware Manufacturing

3326 Spring and Wire Product Manufacturing

33261 Spring and Wire Product Manufacturing

332611 Spring (Heavy Gauge) Manufacturing

332612 Spring (Light Gauge) Manufacturing

332618 Other Fabricated Wire Product Manufacturing

3327 Machine Shops; Turned Product; and Screw, Nut and Bolt Manufacturing

33271 Machine Shops

332710 Machine Shops

33272 Turned Product and Screw, Nut and Bolt Manufacturing

332721 Precision Turned Product Manufacturing

332722 Bolt, Nut, Screw, Rivet, and Washer Manufacturing

3328 Coating, Engraving, Heat Treating, and Allied Activities

33281 Coating, Engraving, Heat Treating, and Allied Activities

332811 Metal Heat Treating

332812 Metal Coating, Engraving (except Jewelry and Silverware), and Allied Services to Manufacturers

332813 Electroplating, Plating, Polishing, Anodizing, and Coloring

3329 Other Fabricated Metal Product Manufacturing

33291 Metal Valve Manufacturing

332911 Industrial Valve Manufacturing

332912 Fluid Power Valve and Hose Fitting Manufacturing

332913 Plumbing Fixture Fitting and Trim Manufacturing

332919 Other Metal Valve and Pipe Fitting Manufacturing

33299 All Other Fabricated Metal Product Manufacturing

332991 Ball and Roller Bearing Manufacturing

332992 Small Arms Ammunition Manufacturing

332993 Ammunition (except Small Arms) Manufacturing

Table I. NAICS Codes

332994	Small Arms Manufacturing
332995	Other Ordnance and Accessories Manufacturing
332996	Fabricated Pipe and Pipe Fitting Manufacturing
332997	Industrial Pattern Manufacturing
332998	Enameled Iron and Metal Sanitary Ware Manufacturing
332999	All Other Miscellaneous Fabricated Metal Product Manufacturing

333 Machinery Manufacturing

3331 Agriculture, Construction, and Mining Machinery Manufacturing

33311 Agricultural Implement Manufacturing

333111	Farm Machinery and Equipment Manufacturing
333112	Lawn and Garden Tractor and Home Lawn and Garden Equipment Manufacturing

33312 Construction Machinery Manufacturing

333120	Construction Machinery Manufacturing

33313 Mining and Oil and Gas Field Machinery Manufacturing

333131	Mining Machinery and Equipment Manufacturing
333132	Oil and Gas Field Machinery and Equipment Manufacturing

3332 Industrial Machinery Manufacturing

33321 Sawmill and Woodworking Machinery Manufacturing

333210	Sawmill and Woodworking Machinery Manufacturing

33322 Plastics and Rubber Industry Machinery Manufacturing

333220	Plastics and Rubber Industry Machinery Manufacturing

33329 Other Industrial Machinery Manufacturing

333291	Paper Industry Machinery Manufacturing
333292	Textile Machinery Manufacturing
333293	Printing Machinery and Equipment Manufacturing
333294	Food Product Machinery Manufacturing
333295	Semiconductor Machinery Manufacturing
333298	All Other Industrial Machinery Manufacturing

3333 Commercial and Service Industry Machinery Manufacturing

33331 Commercial and Service Industry Machinery Manufacturing

333311	Automatic Vending Machine Manufacturing
333312	Commercial Laundry, Drycleaning, and Pressing Machine Manufacturing
333313	Office Machinery Manufacturing
333314	Optical Instrument and Lens Manufacturing
333315	Photographic and Photocopying Equipment Manufacturing
333319	Other Commercial and Service Industry Machinery Manufacturing

3334 Ventilation, Heating, Air-Conditioning, and Commercial Refrigeration

33341 Equipment Manufacturing Ventilation, Heating, Air-Conditioning, and Commercial Refrigeration Equipment Manufacturing

333411	Air Purification Equipment Manufacturing
333412	Industrial and Commercial Fan and Blower Manufacturing

Table I. NAICS Codes

333414	Heating Equipment (except Warm Air Furnaces) Manufacturing
333415	Air-Conditioning and Warm Air Heating Equipment and Commercial and Industrial Refrigeration Equipment Manufacturing

3335 Metalworking Machinery Manufacturing

33351 Metalworking Machinery Manufacturing

333511	Industrial Mold Manufacturing
333512	Machine Tool (Metal Cutting Types) Manufacturing
333513	Machine Tool (Metal Forming Types) Manufacturing
333514	Special Die and Tool, Die Set, Jig, and Fixture Manufacturing
333515	Cutting Tool and Machine Tool Accessory Manufacturing
333516	Rolling Mill Machinery and Equipment Manufacturing
333518	Other Metalworking Machinery Manufacturing

3336 Engine, Turbine, and Power Transmission Equipment Manufacturing

33361 Engine, Turbine, and Power Transmission Equipment Manufacturing

333611	Turbine and Turbine Generator Set Units Manufacturing
333612	Speed Changer, Industrial High-Speed Drive, and Gear Manufacturing
333613	Mechanical Power Transmission Equipment Manufacturing
333618	Other Engine Equipment Manufacturing

3339 Other General Purpose Machinery Manufacturing

33391 Pump and Compressor Manufacturing

333911	Pump and Pumping Equipment Manufacturing
333912	Air and Gas Compressor Manufacturing
333913	Measuring and Dispensing Pump Manufacturing

33392 Material Handling Equipment Manufacturing

333921	Elevator and Moving Stairway Manufacturing
333922	Conveyor and Conveying Equipment Manufacturing
333923	Overhead Traveling Crane, Hoist, and Monorail System Manufacturing
333924	Industrial Truck, Tractor, Trailer, and Stacker Machinery Manufacturing

33399 All Other General Purpose Machinery Manufacturing

333991	Power-Driven Handtool Manufacturing
333992	Welding and Soldering Equipment Manufacturing
333993	Packaging Machinery Manufacturing
333994	Industrial Process Furnace and Oven Manufacturing
333995	Fluid Power Cylinder and Actuator Manufacturing
333996	Fluid Power Pump and Motor Manufacturing
333997	Scale and Balance Manufacturing
333999	All Other Miscellaneous General Purpose Machinery Manufacturing

334 Computer and Electronic Product Manufacturing

3341 Computer and Peripheral Equipment Manufacturing

33411 Computer and Peripheral Equipment Manufacturing

334111	Electronic Computer Manufacturing
334112	Computer Storage Device Manufacturing
334113	Computer Terminal Manufacturing
334119	Other Computer Peripheral Equipment Manufacturing

Table I. NAICS Codes

3342 Communications Equipment Manufacturing

33421 Telephone Apparatus Manufacturing

334210 Telephone Apparatus Manufacturing

33422 Radio and Television Broadcasting and Wireless Communications Equipment Manufacturing

334220 Radio and Television Broadcasting and Wireless Communications Equipment Manufacturing

33429 Other Communications Equipment Manufacturing

334290 Other Communications Equipment Manufacturing

3343 Audio and Video Equipment Manufacturing

33431 Audio and Video Equipment Manufacturing

334310 Audio and Video Equipment Manufacturing

3344 Semiconductor and Other Electronic Component Manufacturing

33441 Semiconductor and Other Electronic Component Manufacturing

334411 Electron Tube Manufacturing

334412 Bare Printed Circuit Board Manufacturing

334413 Semiconductor and Related Device Manufacturing

334414 Electronic Capacitor Manufacturing

334415 Electronic Resistor Manufacturing

334416 Electronic Coil, Transformer, and Other Inductor Manufacturing

334417 Electronic Connector Manufacturing

334418 Printed Circuit Assembly (Electronic Assembly) Manufacturing

334419 Other Electronic Component Manufacturing

3345 Navigational, Measuring, Electromedical, and Control Instruments Manufacturing

33451 Navigational, Measuring, Electromedical, and Control Instruments Manufacturing

334510 Electromedical and Electrotherapeutic Apparatus Manufacturing

334511 Search, Detection, Navigation, Guidance, Aeronautical, and Nautical System and Instrument Manufacturing

334512 Automatic Environmental Control Manufacturing for Residential, Commercial, and Appliance Use

334513 Instruments and Related Products Manufacturing for Measuring, Displaying, and Controlling Industrial Process Variables

334514 Totalizing Fluid Meter and Counting Device Manufacturing

334515 Instrument Manufacturing for Measuring and Testing Electricity and Electrical Signals

334516 Analytical Laboratory Instrument Manufacturing

334517 Irradiation Apparatus Manufacturing

334518 Watch, Clock, and Part Manufacturing

334519 Other Measuring and Controlling Device Manufacturing

3346 Manufacturing and Reproducing Magnetic and Optical Media

33461 Manufacturing and Reproducing Magnetic and Optical Media

334612 Prerecorded Compact Disc (except Software), Tape, and Record Reproducing (except facilities primarily engaged in mass reproducing pre-recorded Video Cassettes, and mass reproducing Video tape or disk)

334613 Magnetic and Optical Recording Media Manufacturing

Table I. NAICS Codes

335	**Electrical Equipment, Appliance, and Component Manufacturing**

3351	**Electric Lighting Equipment Manufacturing**

33511	**Electric Lamp Bulb and Part Manufacturing**
335110	Electric Lamp Bulb and Part Manufacturing

33512	**Lighting Fixture Manufacturing**
335121	Residential Electric Lighting Fixture Manufacturing
335122	Commercial, Industrial, and Institutional Electric Lighting Fixture Manufacturing
335129	Other Lighting Equipment Manufacturing

3352	**Household Appliance Manufacturing**

33521	**Small Electrical Appliance Manufacturing**
335211	Electric Housewares and Household Fan Manufacturing
335212	Household Vacuum Cleaner Manufacturing

33522	**Major Appliance Manufacturing**
335221	Household Cooking Appliance Manufacturing
335222	Household Refrigerator and Home Freezer Manufacturing
335224	Household Laundry Equipment Manufacturing
335228	Other Major Household Appliance Manufacturing

3353	**Electrical Equipment Manufacturing**

33531	**Electrical Equipment Manufacturing**
335311	Power, Distribution, and Specialty Transformer Manufacturing
335312	Motor and Generator Manufacturing (except facilities primarily engaged in armature rewinding on a factory basis)
335313	Switchgear and Switchboard Apparatus Manufacturing
335314	Relay and Industrial Control Manufacturing

3359	**Other Electrical Equipment and Component Manufacturing**

33591	**Battery Manufacturing**
335911	Storage Battery Manufacturing
335912	Primary Battery Manufacturing

33592	**Communication and Energy Wire and Cable Manufacturing**
335921	Fiber Optic Cable Manufacturing
335929	Other Communication and Energy Wire Manufacturing

33593	**Wiring Device Manufacturing**
335931	Current-Carrying Wiring Device Manufacturing
335932	Noncurrent-Carrying Wiring Device Manufacturing

33599	**All Other Electrical Equipment and Component Manufacturing**
335991	Carbon and Graphite Product Manufacturing
335999	All Other Miscellaneous Electrical Equipment and Component Manufacturing

336	**Transportation Equipment Manufacturing**

3361	**Motor Vehicle Manufacturing**

33611	**Automobile and Light Duty Motor Vehicle Manufacturing**
336111	Automobile Manufacturing
336112	Light Truck and Utility Vehicle Manufacturing

Table I. NAICS Codes

33612	**Heavy Duty Truck Manufacturing**
336120	Heavy Duty Truck Manufacturing
3362	**Motor Vehicle Body and Trailer Manufacturing**
33621	**Motor Vehicle Body and Trailer Manufacturing**
336211	Motor Vehicle Body Manufacturing
336212	Truck Trailer Manufacturing
336213	Motor Home Manufacturing
336214	Travel Trailer and Camper Manufacturing
3363	**Motor Vehicle Parts Manufacturing**
33631	**Motor Vehicle Gasoline Engine and Engine Parts Manufacturing**
336311	Carburetor, Piston, Piston Ring, and Valve Manufacturing
336312	Gasoline Engine and Engine Parts Manufacturing
33632	**Motor Vehicle Electrical and Electronic Equipment Manufacturing**
336321	Vehicular Lighting Equipment Manufacturing
336322	Other Motor Vehicle Electrical and Electronic Equipment Manufacturing
33633	**Motor Vehicle Steering and Suspension Components (except Spring) Manufacturing**
336330	Motor Vehicle Steering and Suspension Components (except Spring) Manufacturing
33634	**Motor Vehicle Brake System Manufacturing**
336340	Motor Vehicle Brake System Manufacturing

33635	**Motor Vehicle Transmission and Power Train Parts Manufacturing**
336350	Motor Vehicle Transmission and Power Train Parts Manufacturing
33636	**Motor Vehicle Seating and Interior Trim Manufacturing**
336360	Motor Vehicle Seating and Interior Trim Manufacturing
33637	**Motor Vehicle Metal Stamping**
336370	Motor Vehicle Metal Stamping
33639	**Other Motor Vehicle Parts Manufacturing**
336391	Motor Vehicle Air-Conditioning Manufacturing
336399	All Other Motor Vehicle Parts Manufacturing
3364	**Aerospace Product and Parts Manufacturing**
33641	**Aerospace Product and Parts Manufacturing**
336411	Aircraft Manufacturing
336412	Aircraft Engine and Engine Parts Manufacturing
336413	Other Aircraft Parts and Auxiliary Equipment Manufacturing
336414	Guided Missile and Space Vehicle Manufacturing
336415	Guided Missile and Space Vehicle Propulsion Unit and Propulsion Unit Parts Manufacturing
336419	Other Guided Missile and Space Vehicle Parts and Auxiliary Equipment Manufacturing
3365	**Railroad Rolling Stock Manufacturing**
33651	**Railroad Rolling Stock Manufacturing**
336510	Railroad Rolling Stock Manufacturing
3366	**Ship and Boat Building**

Table I. NAICS Codes

33661	**Ship and Boat Building**
336611	Ship Building and Repairing
336612	Boat Building
3369	**Other Transportation Equipment Manufacturing**
33699	**Other Transportation Equipment Manufacturing**
336991	Motorcycle, Bicycle, and Parts Manufacturing
336992	Military Armored Vehicle, Tank, and Tank Component Manufacturing
336999	All Other Transportation Equipment Manufacturing
337	**Furniture and Related Product Manufacturing**
3371	**Household and Institutional Furniture and Kitchen Cabinet Manufacturing**
33711	**Wood Kitchen Cabinet and Countertop Manufacturing**
337110	Wood Kitchen Cabinet and Countertop Manufacturing (except facilities primarily engaged in the retail sale of household furniture and that manufacture custom wood kitchen cabinets and counter tops)
33712	**Household and Institutional Furniture Manufacturing**
337121	Upholstered Household Furniture Manufacturing (except facilities primarily engaged in the retail sale of household furniture and that manufacture custom made upholstered household furniture)
337122	Nonupholstered Wood Household Furniture Manufacturing (except facilities primarily engaged in the retail sale of household furniture and that manufacture nonupholstered, household type, custom wood furniture)
337124	Metal Household Furniture Manufacturing
337125	Household Furniture (except Wood and Metal) Manufacturing
337127	Institutional Furniture Manufacturing

337129	Wood Television, Radio, and Sewing Machine Cabinet Manufacturing
3372	**Office Furniture (including Fixtures)Manufacturing**
33721	**Office Furniture (including Fixtures)Manufacturing**
337211	Wood Office Furniture Manufacturing
337212	Custom Architectural Woodwork and Millwork Manufacturing
337214	Office Furniture (except Wood) Manufacturing
337215	Showcase, Partition, Shelving, and Locker Manufacturing
3379	**Other Furniture Related Product Manufacturing**
33791	**Mattress Manufacturing**
337910	Mattress Manufacturing
33792	**Blind and Shade Manufacturing**
337920	Blind and Shade Manufacturing
339	**Miscellaneous Manufacturing**
3391	**Medical Equipment and Supplies Manufacturing**
33911	**Medical Equipment and Supplies Manufacturing**
339111	Laboratory Apparatus and Furniture Manuf.
339112	Surgical and Medical Instrument Manufacturing
339113	Surgical Appliance and Supplies Manufacturing (except facilities primarily engaged in manufacturing orthopedic devices to prescription in a retail environment)
339114	Dental Equipment and Supplies Manufacturing
339115	Ophthalmic Goods Manufacturing (except lens grinding facilities that are primarily engaged in the retail sale of eyeglasses and contact lenses to prescription for individuals)

Table I. NAICS Codes

3399	**Other Miscellaneous Manufacturing**

33991 Jewelry and Silverware Manufacturing

339911	Jewelry (except Costume) Manufacturing
339912	Silverware and Hollowware Manufacturing
339913	Jewelers' Material and Lapidary Work Manufacturing
339914	Costume Jewelry and Novelty Manufacturing

33992 Sporting and Athletic Goods Manufacturing

| 339920 | Sporting and Athletic Goods Manufacturing |

33993 Doll, Toy, and Game Manufacturing

| 339931 | Doll and Stuffed Toy Manufacturing |
| 339932 | Game, Toy, and Children's Vehicle Manufacturing |

33994 Office Supplies (except Paper) Manufacturing

339941	Pen and Mechanical Pencil Manufacturing
339942	Lead Pencil and Art Good Manufacturing
339943	Marking Device Manufacturing
339944	Carbon Paper and Inked Ribbon Manufacturing

33995 Sign Manufacturing

| 339950 | Sign Manufacturing |

33999 All Other Miscellaneous Manufacturing

339991	Gasket, Packing, and Sealing Device Manufacturing
339992	Musical Instrument Manufacturing
339993	Fastener, Button, Needle, and Pin Manufacturing
339994	Broom, Brush, and Mop Manufacturing
339995	Burial Casket Manufacturing
339999	All Other Miscellaneous Manufacturing
113310	**Logging**

111998	**All Other Miscellaneous Crop Farming** (Limited to facilities primarily engaged in reducing maple sap to maple syrup)
211112	**Natural Gas Liquid Extraction** (limited to facilities that recover sulfur from natural gas)
212324	**Kaolin and Ball Clay Mining** (limited to facilities operating without a mine or quarry and that are primarily engaged in beneficiating kaolin and clay)
212325	**Clay and Ceramic and Refractory Minerals Mining** (limited to facilities operating without a mine or quarry and that are primarily engaged in beneficiating clay and ceramic and refractory minerals)
212393	**Other Chemical and Fertilizer Mineral Mining** (limited to facilities operating without a mine or quarry that are primarily engaged in beneficiating chemical or fertilizer mineral raw materials)
212399	**All Other Nonmetallic Mineral Mining** (limited to facilities operating without a mine or quarry that are primarily engaged in beneficiating nonmetallic minerals)
488390	**Other Support Activities for Water Transportation** (limited to facilities that are primarily engaged in providing routine repair and maintenance of ships and boats from floating drydocks)
511110	**Newspaper Publishers**
511120	**Periodical Publishers**
511130	**Book Publishers**
511140	**Directory and Mailing List Publishers** (except Facilities that are primarily engaged in furnishing services for direct mail advertising including address list compilers, address list publishers, address list publishers and printing combined, address list publishing, business directory publishers, catalog of collections publishers, catalog of collections publishers and printing combined, mailing list compilers, directory compilers, and mailing list compiling services)
511191	**Greeting Card Publishers**
511199	**All Other Publishers**
512220	**Integrated Record Production/Distribution**

Table I. NAICS Codes

512230	**Music Publishers** (except facilities primarily Engaged in Music copyright authorizing use, Music copyright buying and licensing, and Music publishers working on their own account)
516110	**Internet Publishing and Broadcasting and web search portals** (limited to facilities primarily engaged in Internet newspaper publishing, Internet periodical publishing, internet book publishing, Miscellaneous Internet publishing, Internet greeting card publishers except web search portals
541710	**Research and Development in the Physical, Engineering, and Life Sciences except Biotechnology** (limited to facilities that are primarily engaged in Guided missile and space vehicle engine research and development, and in Guided missile and space vehicle parts (except engines) research and development)
811490	**Other Personal and Household Goods Repair and Maintenance (limited to facilities that are primarily engaged in repairing and servicing pleasure and sail boats without retailing new boats (previously classified under SIC 3732, Boat Building and Repairing (pleasure boat building)**

Table II

1.2 NAICS codes that correspond to SIC codes other than 20 through 39:

212	Mining (except Oil and Gas)
2121	**Coal Mining**
212111	Bituminous Coal and Lignite Surface Mining
212112	Bituminous Coal Underground Mining
212113	Anthracite Mining
2122	**Metal Ore Mining**
212221	Gold Ore Mining
212222	Silver Ore Mining
212231	Lead Ore and Zinc Ore Mining
212234	Copper Ore and Nickel Ore Mining
212299	All Other Metal Ore Mining
221	**Utilities**
22111	**Electric Power Generation (limited to facilities that combust coal and/or oil for the purpose of generating power for distribution in commerce)**
221111	Hydroelectric Power Generation
221112	Fossil Fuel Electric Power Generation
221113	Nuclear Electric Power Generation
221119	Other Electric Power Generation
221121	Electric Bulk Power Transmission and Control
221122	Electric Power Distribution

221330 **Steam and Air Conditioning Supply** Limited to facilities engaged in providing combinations of electric, gas and other services, not elsewhere classified (NEC) (previously classified under SIC 4939, Combination Utility Services Not Elsewhere Classified.)

424690 **Other Chemical and Allied Products Merchant Wholesalers**

424710 **Petroleum Bulk Stations and Terminals**

425110 **Business to Business Electronic Markets** (limited to facilities previously classified in 5169, Chemicals and Allied Products, NEC)

425120 **Wholesale Trade Agents and Brokers** (limited to facilities previously classified in 5169, Chemicals and Allied Products, NEC)

562112 **Hazardous Waste Collection** (limited to facilities primarily engaged in solvent recovery services on a contract or fee basis)

562211 **Hazardous Waste Treatment and Disposal** (limited to facilities regulated under the Resource Conservation and Recovery Act, subtitle C, 42 U.S.C. 6921, *et seq.*)

562212 **Solid Waste Landfill** (limited to facilities regulated under the Resource Conservation and Recovery Act, subtitle C, 42 U.S.C. 6921, *et seq.*)

562213 **Solid Waste Combustors and Incinerators** (Limited to facilities regulated under the Resource Conservation and Recovery Act, subtitle C, 42 U.S.C. 6921 *et seq.*)

562219 **Other Nonhazardous Waste Treatment and Disposal** (Limited to facilities regulated under the Resource Conservation and Recovery Act, subtitle C, 42 U.S.C. 6921 *et seq.*)

562920 **Materials Recovery Facilities** (Limited to facilities regulated under the Resource Conservation and Recovery Act, subtitle C, 42 U.S.C. 6921 *et seq.*)

Table II. EPCRA Section 313 Chemical List For Reporting Year 2012
(including Toxic Chemical Categories)

Individually listed EPCRA Section 313 chemicals with CAS numbers are arranged alphabetically starting on page II-3. Following the alphabetical list, the EPCRA Section 313 chemicals are arranged in CAS number order. Covered chemical categories follow.

Certain EPCRA Section 313 chemicals listed in Table II have parenthetic "qualifiers." These qualifiers indicate that these EPCRA Section 313 chemicals are subject to the section 313 reporting requirements if manufactured, processed, or otherwise used in a specific form or when a certain activity is performed. The following chemicals are reportable only if they are manufactured, processed, or otherwise used in the specific form(s) listed below:

Chemical/ Chemical Category	CAS Number	Qualifier
Aluminum (fume or dust)	7429-90-5	<u>Only</u> if it is a fume or dust form.
Aluminum oxide (fibrous forms)	1344-28-1	<u>Only</u> if it is a fibrous form.
Ammonia (includes anhydrous ammonia and aqueous ammonia from water dissociable ammonium salts and other sources; 10 percent of total aqueous ammonia is reportable under this listing)	7664-41-7	<u>Only</u> 10% of aqueous forms. 100% of anhydrous forms.
Asbestos (friable)	1332-21-4	<u>Only</u> if it is a friable form.
Hydrochloric acid (acid aerosols including mists, vapors, gas, fog, and other airborne forms of any particle size)	7647-01-0	<u>Only</u> if it is an aerosol form as defined.
Nitrate compounds (water dissociable; reportable only when in aqueous solution)	NA	<u>Only</u> if in aqueous solution
Phosphorus (yellow or white)	7723-14-0	<u>Only</u> if it is a yellow or white form.
Sulfuric acid (acid aerosols including mists, vapors, gas, fog, and other airborne forms of any particle size)	7664-93-9	<u>Only</u> if it is an aerosol form as defined.
Vanadium (except when contained in an alloy)	7440-62-2	<u>Except</u> if it is contained in an alloy.
Zinc (fume or dust)	7440-66-6	<u>Oly</u> if it is in a fume or dust form.

The qualifier for the following three chemicals is based on the chemical activity rather than the form of the chemical. These chemicals are subject to EPCRA section 313 reporting requirements only when the indicated activity is performed.

Chemical/ Chemical Category	CAS Number	Qualifier
Dioxin and dioxin-like compounds (manufacturing; and the processing or otherwise use of dioxin and dioxin-like compounds if the dioxin and dioxin-like compounds are present as contaminants in a chemical and if they were created during the manufacture of that chemical.)	NA	<u>Only</u> if they are manufactured at the facility; or are processed or otherwise used when present as contaminants in a chemical but only if they were created during the manufacture of that chemical.
Isopropyl alcohol (only persons who manufacture by the strong acid process are subject, no supplier notification)	67-63-0	<u>Only</u> if it is being manufactured by the strong acid process. Facilities that process or otherwise use isopropyl alcohol are <u>not</u> covered and should <u>not</u> file a report.
Saccharin (only persons who manufacture are subject, no supplier notification)	81-07-2	<u>Only</u> if it is being manufactured.

There are no supplier notification requirements for isopropyl alcohol and saccharin since the processors and users of these chemicals are not required to report. Manufacturers of these chemicals do not need to notify their customers that these are reportable EPCRA section 313 chemicals.

Table II. EPCRA Section 313 Chemical List for Reporting Year 2012

Note: Chemicals may be added to or deleted from the list. The Emergency Planning and Community Right-to-Know Call Center will provide up-to-date information on the status of these changes. See section B.3.c of the instructions for more information on the *de minimis* % limits listed below. There are no *de minimis* levels for PBT chemicals since the *de minimis* exemption is not available for these chemicals (an asterisk appears where a *de minimis* limit would otherwise appear in Table II). However, for purposes of the supplier notification requirement only, such limits are provided in Appendix D.

Chemical Qualifiers

This table contains the list of individual EPCRA Section 313 chemicals and categories of chemicals subject to 2009 calendar year reporting. Some of the EPCRA Section 313 chemicals listed have parenthetic qualifiers listed next to them. An EPCRA Section 313 chemical that is listed without a qualifier is subject to reporting in all forms in which it is manufactured, processed, and otherwise used.

Fume or dust. Two of the metals on the list (aluminum and zinc) contain the qualifier "fume or dust." Fume or dust refers to dry forms of these metals but does not refer to "wet" forms such as solutions or slurries. As explained in Section B.3.a of these instructions, the term manufacture includes the generation of an EPCRA Section 313 chemical as a byproduct or impurity. In such cases, a facility should determine if, for example, it generated more than 25,000 pounds of aluminum fume or dust in the reporting year as a result of its activities. If so, the facility must report that it manufactures "aluminum (fume or dust)." Similarly, there may be certain technologies in which one of these metals is processed in the form of a fume or dust to make other EPCRA Section 313 chemicals or other products for distribution in commerce. In reporting releases, the facility would only report releases of the fume or dust.

EPA considers dusts to consist of solid particles generated by any mechanical processing of materials including crushing, grinding, rapid impact, handling, detonation, and decrepitation of organic and inorganic materials such as rock, ore, and metal. Dusts do not tend to flocculate, except under electrostatic forces.

EPA considers a fume to be an airborne dispersion consisting of small solid particles created by condensation from a gaseous state, in distinction to a gas or vapor. Fumes arise from the heating of solids such as lead. The condensation is often accompanied by a chemical reaction, such as oxidation. Fumes flocculate and sometimes coalesce.

Manufacturing qualifiers. Two of the entries in the EPCRA Section 313 chemical list contain a qualifier relating to manufacture. For isopropyl alcohol, the qualifier is "only persons who manufacture by the strong acid process are subject, no supplier notification." For saccharin, the qualifier is "only persons who manufacture are subject, no supplier notification." For isopropyl alcohol, the qualifier means that only facilities manufacturing isopropyl alcohol by the strong acid process are required to report. In the case of saccharin, only manufacturers of the EPCRA Section 313 chemical are subject to the reporting requirements. A facility that only processes or otherwise uses either of these EPCRA Section 313 chemicals is not required to report for these EPCRA Section 313 chemicals. In both cases, supplier notification does not apply because only manufacturers, not users, of these two EPCRA Section 313 chemicals must report.

Ammonia (includes anhydrous ammonia and aqueous ammonia from water dissociable ammonium salts and other sources; 10 percent of total aqueous ammonia is reportable under this listing). The qualifier for ammonia means that anhydrous forms of ammonia are 100% reportable and aqueous forms are limited to 10% of total aqueous ammonia. Therefore when determining threshold and releases and other waste management quantities all anhydrous ammonia is included but only 10% of total aqueous ammonia is included. Any evaporation of ammonia from aqueous ammonia solutions is considered anhydrous ammonia and should be included in threshold determinations and release and other waste management calculations.

Sulfuric acid and Hydrochloric acid (acid aerosols including mists, vapors, gas, fog, and other airborne forms of any particle size). The qualifier for sulfuric acid and hydrochloric acid means that the only forms of these chemicals that are reportable are airborne forms. Aqueous solutions are not covered by this listing but any aerosols generated from aqueous solutions are covered.

Nitrate compounds (water dissociable; reportable only when in aqueous solution). The qualifier for the nitrate compounds category limits the reporting to nitrate compounds that dissociate in water, generating nitrate ion. For the purposes of threshold determinations the entire weight of the nitrate compound must be included in all calculations. For the purposes of reporting releases and other waste management quantities only the weight of the nitrate ion should be included in the calculations of these quantities.

Phosphorus (yellow or white). The listing for phosphorus is qualified by the term "yellow or white." This means that only manufacturing, processing, or otherwise use of phosphorus in the yellow or white chemical form triggers reporting. Conversely, manufacturing, processing, or otherwise use of "black" or "red" phosphorus does not trigger reporting. Supplier notification also applies only to distribution of yellow or white phosphorus.

Asbestos (friable). The listing for asbestos is qualified by the term "friable," referring to the physical characteristic of being able to be crumbled, pulverized, or reducible to a powder with hand pressure. Only manufacturing, processing, or otherwise use of asbestos in the friable form triggers reporting. Supplier notification applies only to distribution of mixtures or other trade name products containing friable asbestos.

Aluminum Oxide (fibrous forms). The listing for aluminum

Table II. EPCRA Section 313 Chemical List for Reporting Year 2012

oxide is qualified by the term "fibrous forms." Fibrous refers to a man-made form of aluminum oxide that is processed to produce strands or filaments which can be cut to various lengths depending on the application. Only manufacturing, processing, or otherwise use of aluminum oxide in the fibrous form triggers reporting. Supplier notification applies only to distribution of mixtures or other trade name products containing fibrous forms of aluminum oxide.

Notes for Sections A and B of following list of TRI chemicals:

"Color Index" indicated by "C.I."

* There are no *de minimis* levels for PBT chemicals, except for supplier notification purposes (see Appendix D).

---

a.   Individually-Listed Toxic Chemicals Arranged Alphabetically

CAS Number	Chemical Name	*De minimis* % Limit
71751-41-2	Abamectin [Avermectin B1]	1.0
30560-19-1	Acephate (Acetylphosphoramidothioic acid O,S-dimethyl ester)	1.0
75-07-0	Acetaldehyde	0.1
60-35-5	Acetamide	0.1
75-05-8	Acetonitrile	1.0
98-86-2	Acetophenone	1.0
53-96-3	2-Acetylaminofluorene	0.1
62476-59-9	Acifluorfen, sodium salt [5-(2-Chloro-4-(trifluoromethyl)phenoxy)-2-nitrobenzoic acid, sodium salt]	1.0
107-02-8	Acrolein	1.0
79-06-1	Acrylamide	0.1
79-10-7	Acrylic acid	1.0
107-13-1	Acrylonitrile	0.1
15972-60-8	Alachlor	1.0
116-06-3	Aldicarb	1.0
309-00-2	Aldrin [1,4:5,8-Dimethanonaphthalene, 1,2,3,4,10,10-hexachloro-1,4,4a,5,8,8a-hexahydro-(1.alpha.,4.alpha.,4a.beta., 5.alpha.,8.alpha.,8a.beta.)-]	*
28057-48-9	d-trans-Allethrin [d-trans-Chrysanthemic acid of d-allethrone]	1.0
107-18-6	Allyl alcohol	1.0
107-11-9	Allylamine	1.0
107-05-1	Allyl chloride	1.0
7429-90-5	Aluminum (fume or dust)	1.0
20859-73-8	Aluminum phosphide	1.0
1344-28-1	Aluminum oxide (fibrous forms)	1.0
834-12-8	Ametryn (N-Ethyl-N=-(1-methylethyl)-6-(methylthio)-1,3,5,-triazine-2,4-diamine)	1.0
117-79-3	2-Aminoanthraquinone	0.1
60-09-3	4-Aminoazobenzene	0.1
92-67-1	4-Aminobiphenyl	0.1

CAS Number	Chemical Name	*De minimis* % Limit
82-28-0	1-Amino-2-methylanthraquinone	0.1
81-49-2	1-Amino-2,4-dibromoanthraquinone	0.1
33089-61-1	Amitraz	1.0
61-82-5	Amitrole	0.1
7664-41-7	Ammonia (includes anhydrous ammonia and aqueous ammonia from water dissociable ammonium salts and other sources; 10 percent of total aqueous ammonia is reportable under this listing)	1.0
101-05-3	Anilazine [4,6-Dichloro-N-(2-chlorophenyl)-1,3,5-triazin-2-amine]	1.0
62-53-3	Aniline	1.0
90-04-0	o-Anisidine	0.1
104-94-9	p-Anisidine	1.0
134-29-2	o-Anisidine hydrochloride	0.1
120-12-7	Anthracene	1.0
7440-36-0	Antimony	1.0
7440-38-2	Arsenic	0.1
1332-21-4	Asbestos (friable)	0.1
1912-24-9	Atrazine (6-Chloro-N-ethyl-N=-(1-methylethyl)-1,3,5-triazine-2,4-diamine)	1.0
7440-39-3	Barium	1.0
22781-23-3	Bendiocarb [2,2-Dimethyl-1,3-benzodioxol-4-ol methylcarbamate]	1.0
1861-40-1	Benfluralin (N-Butyl-N-ethyl-2,6-dinitro-4-(trifluoromethyl)benzenamine)	1.0
17804-35-2	Benomyl	1.0
98-87-3	Benzal chloride	1.0
55-21-0	Benzamide	1.0
71-43-2	Benzene	0.1
92-87-5	Benzidine	0.1
98-07-7	Benzoic trichloride (Benzotrichloride)	0.1
191-24-2	Benzo(g,h,i)perylene	*
98-88-4	Benzoyl chloride	1.0
94-36-0	Benzoyl peroxide	1.0
100-44-7	Benzyl chloride	1.0
7440-41-7	Beryllium	0.1
82657-04-3	Bifenthrin	1.0
92-52-4	Biphenyl	1.0
3296-90-0	2,2-bis(Bromomethyl)-1,3-propanediol	0.1
111-91-1	Bis(2-chloroethoxy) methane	1.0
111-44-4	Bis(2-chloroethyl) ether	1.0
542-88-1	Bis(chloromethyl) ether	0.1
108-60-1	Bis(2-chloro-1-methylethyl)ether	1.0
56-35-9	Bis(tributyltin) oxide	1.0
10294-34-5	Boron trichloride	1.0
7637-07-2	Boron trifluoride	1.0
314-40-9	Bromacil (5-Bromo-6-methyl-3-(1-methylpropyl)-2,4(1H,3H)-pyrimidinedione)	1.0
53404-19-6	Bromacil, lithium salt [2,4(1H,3H)-Pyrimidinedione,5-bromo-6-	1.0

Table II. EPCRA Section 313 Chemical List for Reporting Year 2012

CAS Number	Chemical Name	*De minimis* % Limit	CAS Number	Chemical Name	*De minimis* % Limit
	methyl-3-(1-methylpropyl), lithium salt]			1-azoniaadamantane chloride	
7726-95-6	Bromine	1.0	106-47-8	p-Chloroaniline	0.1
			108-90-7	Chlorobenzene	1.0
35691-65-7	1-Bromo-1-(bromomethyl)- 1,3-propanedicarbonitrile	1.0	510-15-6	Chlorobenzilate [Benzeneacetic acid, 4-chloro-.alpha.- (4-chlorophenyl)-.alpha.-hydroxy-, ethyl ester]	1.0
353-59-3	Bromochlorodifluoromethane (Halon 1211)	1.0	75-68-3	1-Chloro-1,1-difluoroethane (HCFC-142b)	1.0
75-25-2	Bromoform (Tribromomethane)	1.0	75-45-6	Chlorodifluoromethane (HCFC-22)	1.0
74-83-9	Bromomethane (Methyl bromide)	1.0	75-00-3	Chloroethane (Ethyl chloride)	1.0
75-63-8	Bromotrifluoromethane (Halon 1301)	1.0	67-66-3	Chloroform	0.1
			74-87-3	Chloromethane (Methyl chloride)	1.0
1689-84-5	Bromoxynil (3,5-Dibromo-4-hydroxybenzonitrile)	1.0	107-30-2	Chloromethyl methyl ether	0.1
			563-47-3	3-Chloro-2-methyl-1-propene	0.1
1689-99-2	Bromoxynil octanoate (Octanoic acid, 2,6-dibromo-4-cyanophenylester)	1.0	104-12-1	p-Chlorophenyl isocyanate	1.0
			76-06-2	Chloropicrin	1.0
357-57-3	Brucine	1.0	126-99-8	Chloroprene	0.1
106-99-0	1,3-Butadiene	0.1	542-76-7	3-Chloropropionitrile	1.0
141-32-2	Butyl acrylate	1.0	63938-10-3	Chlorotetrafluoroethane	1.0
71-36-3	n-Butyl alcohol	1.0	354-25-6	1-Chloro-1,1,2,2-tetrafluoroethane (HCFC-124a)	1.0
78-92-2	sec-Butyl alcohol	1.0			
75-65-0	tert-Butyl alcohol	1.0	2837-89-0	2-Chloro-1,1,1,2-tetrafluoroethane (HCFC-124)	1.0
106-88-7	1,2-Butylene oxide	0.1			
123-72-8	Butyraldehyde	1.0	1897-45-6	Chlorothalonil [1,3-Benzenedicarbonitrile, 2,4,5,6-tetrachloro-]	0.1
7440-43-9	Cadmium	0.1			
156-62-7	Calcium cyanamide	1.0			
133-06-2	Captan [1H-Isoindole-1,3(2H)-dione, 3a,4,7,7a-tetrahydro-2-[(trichloromethyl)thio]-]	1.0	95-69-2	p-Chloro-o-toluidine	0.1
			75-88-7	2-Chloro-1,1,1-trifluoroethane (HCFC-133a)	1.0
63-25-2	Carbaryl [1-Naphthalenol, methylcarbamate]	1.0	75-72-9	Chlorotrifluoromethane (CFC-13)	1.0
1563-66-2	Carbofuran	1.0	460-35-5	3-Chloro-1,1,1-trifluoropropane (HCFC-253fb)	1.0
75-15-0	Carbon disulfide	1.0			
56-23-5	Carbon tetrachloride	0.1	5598-13-0	Chlorpyrifos methyl [O,O-Dimethyl-O-(3,5,6-trichloro-2-pyridyl)phosphorothioate]	1.0
463-58-1	Carbonyl sulfide	1.0			
5234-68-4	Carboxin (5,6-Dihydro-2-methyl-N-phenyl-1,4-oxathiin-3-carboxamide)	1.0	64902-72-3	Chlorsulfuron [2-Chloro-N-[[(4-methoxy-6-methyl-1,3,5-triazin-2-yl)amino]carbonyl] benzenesulfonamide]	1.0
120-80-9	Catechol	0.1			
2439-01-2	Chinomethionat [6-Methyl-1,3-dithiolo[4,5-b]quinoxalin-2-one]	1.0	7440-47-3	Chromium	1.0
			4680-78-8	C.I. Acid Green 3	1.0
			6459-94-5	C.I. Acid Red 114	0.1
133-90-4	Chloramben [Benzoic acid, 3-amino-2,5-dichloro-]	1.0	569-64-2	C.I. Basic Green 4	1.0
			989-38-8	C.I. Basic Red 1	1.0
57-74-9	Chlordane [4,7-Methanoindan, 1,2,4,5,6,7,8,8-octachloro-2,3,3a,4,7,7a-hexahydro-]	*	1937-37-7	C.I. Direct Black 38	0.1
			2602-46-2	C.I. Direct Blue 6	0.1
			28407-37-6	C.I. Direct Blue 218	1.0
115-28-6	Chlorendic acid	0.1	16071-86-6	C.I. Direct Brown 95	0.1
90982-32-4	Chlorimuron ethyl [Ethyl-2-[[[[(4-chloro-6-methoxyprimidin-2-yl)amino]carbonyl]amino]sulfonyl] benzoate]	1.0	2832-40-8	C.I. Disperse Yellow 3	1.0
			3761-53-3	C.I. Food Red 5	0.1
			81-88-9	C.I. Food Red 15	1.0
			3118-97-6	C.I. Solvent Orange 7	1.0
7782-50-5	Chlorine	1.0	97-56-3	C.I. Solvent Yellow 3	0.1
10049-04-4	Chlorine dioxide	1.0	842-07-9	C.I. Solvent Yellow 14	1.0
79-11-8	Chloroacetic acid	1.0			
532-27-4	2-Chloroacetophenone	1.0	492-80-8	C.I. Solvent Yellow 34 (Auramine)	0.1
4080-31-3	1-(3-Chloroallyl)-3,5,7-triaza-	1.0			

Table II. EPCRA Section 313 Chemical List for Reporting Year 2012

CAS Number	Chemical Name	*De minimis* % Limit
128-66-5	C.I. Vat Yellow 4	1.0
7440-48-4	Cobalt	0.1
7440-50-8	Copper	1.0
8001-58-9	Creosote	0.1
120-71-8	p-Cresidine	0.1
108-39-4	m-Cresol	1.0
95-48-7	o-Cresol	1.0
106-44-5	p-Cresol	1.0
1319-77-3	Cresol (mixed isomers)	1.0
4170-30-3	Crotonaldehyde	1.0
98-82-8	Cumene	1.0
80-15-9	Cumene hydroperoxide	1.0
135-20-6	Cupferron [Benzeneamine, N-hydroxy-N-nitroso, ammonium salt]	0.1
21725-46-2	Cyanazine	1.0
1134-23-2	Cycloate	1.0
110-82-7	Cyclohexane	1.0
108-93-0	Cyclohexanol	1.0
68359-37-5	Cyfluthrin [3-(2,2-Dichloroethenyl)-2,2-dimethylcyclopropanecarboxylic acid, cyano(4-fluoro-3-phenoxyphenyl) methyl ester]	1.0
68085-85-8	Cyhalothrin [3-(2-Chloro-3,3,3-trifluoro-1-propenyl)-2,2-dimethylcyclopropane-carboxylic acid cyano(3-phenoxyphenyl)methyl ester]	1.0
94-75-7	2,4-D [Acetic acid, (2,4-dichlorophenoxy)-]	0.1
533-74-4	Dazomet (Tetrahydro-3,5-dimethyl-2H-1,3,5-thiadiazine-2-thione)	1.0
53404-60-7	Dazomet, sodium salt [Tetrahydro-3,5-dimethyl-2H-1,3,5-thiadiazine-2-thione, ion(1-), sodium]	1.0
94-82-6	2,4-DB	1.0
1929-73-3	2,4-D butoxyethyl ester	0.1
94-80-4	2,4-D butyl ester	0.1
2971-38-2	2,4-D chlorocrotyl ester	0.1
1163-19-5	Decabromodiphenyl oxide	1.0
13684-56-5	Desmedipham	1.0
1928-43-4	2,4-D 2-ethylhexyl ester	0.1
53404-37-8	2,4-D 2-ethyl-4-methylpentyl ester	0.1
2303-16-4	Diallate [Carbamothioic acid, bis(1-methylethyl)-S-(2,3-dichloro-2-propenyl) ester]	1.0
615-05-4	2,4-Diaminoanisole	0.1
39156-41-7	2,4-Diaminoanisole sulfate	0.1
101-80-4	4,4'-Diaminodiphenyl ether	0.1
95-80-7	2,4-Diaminotoluene	0.1
25376-45-8	Diaminotoluene (mixed isomers)	0.1
333-41-5	Diazinon	1.0
334-88-3	Diazomethane	1.0
132-64-9	Dibenzofuran	1.0
96-12-8	1,2-Dibromo-3-chloropropane (DBCP)	0.1

CAS Number	Chemical Name	*De minimis* % Limit
106-93-4	1,2-Dibromoethane (Ethylene dibromide)	0.1
124-73-2	Dibromotetrafluoroethane (Halon 2402)	1.0
84-74-2	Dibutyl phthalate	1.0
1918-00-9	Dicamba (3,6-Dichloro-2-methoxybenzoic acid)	1.0
99-30-9	Dichloran [2,6-Dichloro-4-nitroaniline]	1.0
95-50-1	1,2-Dichlorobenzene	1.0
541-73-1	1,3-Dichlorobenzene	1.0
106-46-7	1,4-Dichlorobenzene	0.1
25321-22-6	Dichlorobenzene (mixed isomers)	0.1
91-94-1	3,3'-Dichlorobenzidine	0.1
612-83-9	3,3'-Dichlorobenzidine dihydrochloride	0.1
64969-34-2	3,3'-Dichlorobenzidine sulfate	0.1
75-27-4	Dichlorobromomethane	0.1
764-41-0	1,4-Dichloro-2-butene	1.0
110-57-6	trans-1,4-Dichloro-2-butene	1.0
1649-08-7	1,2-Dichloro-1,1-difluoroethane (HCFC-132b)	1.0
75-71-8	Dichlorodifluoromethane (CFC-12)	1.0
107-06-2	1,2-Dichloroethane (Ethylene dichloride)	0.1
540-59-0	1,2-Dichloroethylene	1.0
1717-00-6	1,1-Dichloro-1-fluoroethane (HCFC-141b)	1.0
75-43-4	Dichlorofluoromethane (HCFC-21)	1.0
75-09-2	Dichloromethane (Methylene chloride)	0.1
127564-92-5	Dichloropentafluoropropane	1.0
13474-88-9	1,1-Dichloro-1,2,2,3,3-pentafluoropropane (HCFC-225cc)	1.0
111512-56-2	1,1-Dichloro-1,2,3,3,3-pentafluoropropane (HCFC-225eb)	1.0
422-44-6	1,2-Dichloro-1,1,2,3,3-pentafluoropropane (HCFC-225bb)	1.0
431-86-7	1,2-Dichloro-1,1,3,3,3-pentafluoropropane (HCFC-225da)	1.0
507-55-1	1,3-Dichloro-1,1,2,2,3-pentafluoropropane (HCFC-225cb)	1.0
136013-79-1	1,3-Dichloro-1,1,2,3,3-pentafluoropropane (HCFC-225ea)	1.0
128903-21-9	2,2-Dichloro-1,1,1,3,3-pentafluoropropane (HCFC-225aa)	1.0
422-48-0	2,3-Dichloro-1,1,1,2,3-pentafluoropropane (HCFC-225ba)	1.0
422-56-0	3,3-Dichloro-1,1,1,2,2-pentafluoropropane (HCFC-225ca)	1.0
97-23-4	Dichlorophene [2,2'-Methylenebis(4-chlorophenol)]	1.0
120-83-2	2,4-Dichlorophenol	1.0
78-87-5	1,2-Dichloropropane	1.0
10061-02-6	trans-1,3-Dichloropropene	0.1
78-88-6	2,3-Dichloropropene	1.0
542-75-6	1,3-Dichloropropylene	0.1

Table II. EPCRA Section 313 Chemical List for Reporting Year 2012

CAS Number	Chemical Name	De minimis % Limit
76-14-2	Dichlorotetrafluoroethane (CFC-114)	1.0
34077-87-7	Dichlorotrifluoroethane	1.0
90454-18-5	Dichloro-1,1,2-trifluoroethane	1.0
812-04-4	1,1-Dichloro-1,2,2-trifluoroethane (HCFC-123b)	1.0
354-23-4	1,2-Dichloro-1,1,2-trifluoroethane (HCFC-123a)	1.0
306-83-2	2,2-Dichloro-1,1,1-trifluoroethane (HCFC-123)	1.0
62-73-7	Dichlorvos [Phosphoric acid, 2,2-dichloroethenyl dimethyl ester]	0.1
51338-27-3	Diclofop methyl [2-[4-(2,4-Dichlorophenoxy)phenoxy] propanoic acid, methyl ester]	1.0
115-32-2	Dicofol [Benzenemethanol, 4-chloro-	1.0
77-73-6	Dicyclopentadiene	1.0
1464-53-5	Diepoxybutane	0.1
111-42-2	Diethanolamine	1.0
38727-55-8	Diethatyl ethyl	1.0
117-81-7	Di(2-ethylhexyl) phthalate (DEHP)	0.1
64-67-5	Diethyl sulfate	0.1
35367-38-5	Diflubenzuron	1.0
101-90-6	Diglycidyl resorcinol ether	0.1
94-58-6	Dihydrosafrole	0.1
55290-64-7	Dimethipin [2,3-Dihydro-5,6-dimethyl-1,4-dithiin 1,1,4,4-tetraoxide]	1.0
60-51-5	Dimethoate	1.0
119-90-4	3,3'-Dimethoxybenzidine	0.1
20325-40-0	3,3'-Dimethoxybenzidine dihydrochloride (o-Dianisidine dihydrochloride)	0.1
111984-09-9	3,3'-Dimethoxybenzidine hydrochloride (o-Dianisidine hydrochloride)	0.1
124-40-3	Dimethylamine	1.0
2300-66-5	Dimethylamine dicamba	1.0
60-11-7	4-Dimethylaminoazobenzene	0.1
121-69-7	N,N-Dimethylaniline	1.0
119-93-7	3,3'-Dimethylbenzidine (o-Tolidine)	0.1
612-82-8	3,3'-Dimethylbenzidine dihydrochloride (o-Tolidine dihydrochloride)	0.1
41766-75-0	3,3'-Dimethylbenzidine dihydrofluoride (o-Tolidine dihydrofluoride)	0.1
79-44-7	Dimethylcarbamyl chloride	0.1
2524-03-0	Dimethyl chlorothiophosphate	1.0
68-12-2	N,N-Dimethylformamide	1.0
57-14-7	1,1-Dimethyl hydrazine	0.1
105-67-9	2,4-Dimethylphenol	1.0
131-11-3	Dimethyl phthalate	1.0
77-78-1	Dimethyl sulfate	0.1
99-65-0	m-Dinitrobenzene	1.0
528-29-0	o-Dinitrobenzene	1.0
100-25-4	p-Dinitrobenzene	1.0

CAS Number	Chemical Name	De minimis % Limit
88-85-7	Dinitrobutyl phenol (Dinoseb)	1.0
534-52-1	4,6-Dinitro-o-cresol	1.0
51-28-5	2,4-Dinitrophenol	1.0
121-14-2	2,4-Dinitrotoluene	0.1
606-20-2	2,6-Dinitrotoluene	0.1
25321-14-6	Dinitrotoluene (mixed isomers)	1.0
39300-45-3	Dinocap	1.0
123-91-1	1,4-Dioxane	0.1
957-51-7	Diphenamid	1.0
122-39-4	Diphenylamine	1.0
122-66-7	1,2-Diphenylhydrazine (Hydrazobenzene)	0.1
2164-07-0	Dipotassium endothall [7-Oxabicyclo(2.2.1)heptane-2,3-dicarboxylic acid, dipotassium salt]	1.0
136-45-8	Dipropyl isocinchomeronate	1.0
138-93-2	Disodium cyanodithioimidocarbonate	1.0
94-11-1	2,4-D isopropyl ester	0.1
541-53-7	2,4-Dithiobiuret	1.0
330-54-1	Diuron	1.0
2439-10-3	Dodine [Dodecylguanidine monoacetate]	1.0
120-36-5	2,4-DP	0.1
1320-18-9	2,4-D propylene glycol butyl ether ester	0.1
2702-72-9	2,4-D sodium salt	0.1
106-89-8	Epichlorohydrin	0.1
13194-48-4	Ethoprop [Phosphorodithioic acid O-ethyl S,S-dipropyl ester]	1.0
110-80-5	2-Ethoxyethanol	1.0
140-88-5	Ethyl acrylate	0.1
100-41-4	Ethylbenzene	0.1
541-41-3	Ethyl chloroformate	1.0
759-94-4	Ethyl dipropylthiocarbamate (EPTC)	1.0
74-85-1	Ethylene	1.0
107-21-1	Ethylene glycol	1.0
151-56-4	Ethyleneimine (Aziridine)	0.1
75-21-8	Ethylene oxide	0.1
96-45-7	Ethylene thiourea	0.1
75-34-3	Ethylidene dichloride	1.0
52-85-7	Famphur	1.0
60168-88-9	Fenarimol [.alpha.-(2-Chlorophenyl)-.alpha.-(4-chlorophenyl)-5-pyrimidinemethanol]	1.0
13356-08-6	Fenbutatin oxide (Hexakis(2-methyl-2-phenylpropyl) distannoxane)	1.0
66441-23-4	Fenoxaprop ethyl [2-(4-((6-Chloro-2-benzoxazolylen)oxy)phenoxy)propanoic acid, ethyl ester]	1.0
72490-01-8	Fenoxycarb [[2-(4-Phenoxyphenoxy)ethyl]carbamic acid	1.0

Table II. EPCRA Section 313 Chemical List for Reporting Year 2012

CAS Number	Chemical Name	De minimis % Limit
	ethyl ester]	
39515-41-8	Fenpropathrin [2,2,3,3-Tetramethylcyclopropane carboxylic acid cyano(3-phenoxyphenyl)methyl ester]	1.0
55-38-9	Fenthion [O,O-Dimethyl O-[3-methyl-4-(methylthio)phenyl] ester, phosphorothioic acid]	1.0
51630-58-1	Fenvalerate [4-Chloro-alpha-(1-methylethyl) benzeneacetic acid cyano (3-phenoxyphenyl) methyl ester]	1.0
14484-64-1	Ferbam [Tris(dimethylcarbamodithioato- S,S')iron]	1.0
69806-50-4	Fluazifop butyl [2-[4-[[5-(Trifluoromethyl)-2-pyridinyl]oxy]phenoxy]propanoic acid, butyl ester]	1.0
2164-17-2	Fluometuron [Urea, N,N-dimethyl-N=-[3-(trifluoromethyl)phenyl]-]	1.0
7782-41-4	Fluorine	1.0
51-21-8	Fluorouracil (5-Fluorouracil)	1.0
69409-94-5	Fluvalinate [N-[2-Chloro-4-(trifluoromethyl)phenyl]-DL-valine(+)-cyano(3-phenoxyphenyl)methyl ester]	1.0
133-07-3	Folpet	1.0
72178-02-0	Fomesafen [5-(2-Chloro-4-(trifluoromethyl)phenoxy)-N-methylsulfonyl-2-nitrobenzamide]	1.0
50-00-0	Formaldehyde	0.1
64-18-6	Formic acid	1.0
76-13-1	Freon 113 [Ethane, 1,1,2-trichloro-1,2,2,-trifluoro-]	1.0
110-00-9	Furan	0.1
556-52-5	Glycidol	0.1
76-44-8	Heptachlor [1,4,5,6,7,8,8-Heptachloro-3a, 4,7,7a-tetrahydro-4,7-methano-1H-indene]	*
118-74-1	Hexachlorobenzene	*
87-68-3	Hexachloro-1,3-butadiene	1.0
319-84-6	alpha-Hexachlorocyclohexane	0.1
77-47-4	Hexachlorocyclopentadiene	1.0
67-72-1	Hexachloroethan	0.1
1335-87-1	Hexachloronaphthalene	1.0
70-30-4	Hexachlorophene	1.0
680-31-9	Hexamethylphosphoramide	0.1
110-54-3	n-Hexane	1.0
51235-04-2	Hexazinone	1.0
67485-29-4	Hydramethylnon [Tetrahydro-5,5-dimethyl-2(1H)-pyrimidinone[3-[4-(trifluoromethyl)phenyl]-1-[2-[4-(trifluoromethyl)phenyl]ethenyl]-2-propenylidene]hydrazone]	1.0
302-01-2	Hydrazine	0.1
10034-93-2	Hydrazine sulfate	0.1
7647-01-0	Hydrochloric acid (acid aerosols including mists, vapors, gas, fog, and other airborne forms of any particle size)	1.0
74-90-8	Hydrogen cyanide	1.0
7664-39-3	Hydrogen fluoride	1.0
7783-06-4	Hydrogen sulfide	1.0
123-31-9	Hydroquinone	1.0
35554-44-0	Imazalil [1-[2-(2,4-Dichlorophenyl)-2-(2-propenyloxy)ethyl]-1H-imidazole]	1.0
55406-53-6	3-Iodo-2-propynyl butylcarbamate	1.0
13463-40-6	Iron pentacarbonyl	1.0
78-84-2	Isobutyraldehyde	1.0
465-73-6	Isodrin	*
25311-71-1	Isofenphos[2-[[Ethoxyl[(1-methylethyl)amino]phosphinothioyl]oxy] benzoic acid 1-methylethyl ester]	1.0
78-79-5	Isoprene	0.1
67-63-0	Isopropyl alcohol (only persons who manufacture by the strong acid process are subject, no supplier notification)	1.0
80-05-7	4,4'-Isopropylidenediphenol	1.0
120-58-1	Isosafrole	1.0
77501-63-4	Lactofen [Benzoic acid, 5-[2-Chloro-4-(trifluoromethyl)phenoxy]-2-nitro-, 2-ethoxy-1-methyl-2-oxoethyl ester]	1.0
7439-92-1	Lead (when lead is contained in stainless steel, brass or bronze alloys the *de minimis* level is 0.1)	*
58-89-9	Lindane [Cyclohexane, 1,2,3,4,5,6-hexachloro-, (1.alpha.,2.alpha.,3.beta.,4.alpha.,5.alpha.,6.beta.)-]	0.1
330-55-2	Linuron	1.0
554-13-2	Lithium carbonate	1.0
121-75-5	Malathion	1.0
108-31-6	Maleic anhydride	1.0
109-77-3	Malononitrile	1.0
12427-38-2	Maneb [Carbamodithioic acid, 1,2-ethanediylbis-, manganese complex]	1.0
7439-96-5	Manganese	1.0
93-65-2	Mecoprop	0.1
149-30-4	2-Mercaptobenzothiazole (MBT)	1.0
7439-97-6	Mercury	*
150-50-5	Merphos	1.0
126-98-7	Methacrylonitrile	1.0
137-42-8	Metham sodium (Sodium methyldithiocarbamate)	1.0
67-56-1	Methanol	1.0
20354-26-1	Methazole [2-(3,4-Dichlorophenyl)-4-methyl-1,2,4-	1.0

Table II. EPCRA Section 313 Chemical List for Reporting Year 2012

CAS Number	Chemical Name	De minimis % Limit
	oxadiazolidine-3,5-dione]	
2032-65-7	Methiocarb	1.0
94-74-6	Methoxone	0.1
	((4-Chloro-2-methylphenoxy) acetic acid)	
	(MCPA)	
3653-48-3	Methoxone sodium salt	0.1
	((4-Chloro-2-methylphenoxy) acetate	
	sodium salt)	
72-43-5	Methoxychlor	*
	[Benzene, 1,1'-(2,2,2-	
	trichloroethylidene)bis[4-methoxy-]	
109-86-4	2-Methoxyethanol	1.0
96-33-3	Methyl acrylate	1.0
1634-04-4	Methyl tert-butyl ether	1.0
79-22-1	Methyl chlorocarbonate	1.0
101-14-4	4,4'-Methylenebis(2-chloroaniline) 0.1	
	(MBOCA)	
101-61-1	4,4'-Methylenebis(N,N-dimethyl)	0.1
	benzenamine	
74-95-3	Methylene bromide	1.0
101-77-9	4,4'-Methylenedianiline	0.1
93-15-2	Methyleugenol	0.1
60-34-4	Methyl hydrazine	1.0
74-88-4	Methyl iodide	1.0
108-10-1	Methyl isobutyl ketone	1.0
624-83-9	Methyl isocyanate	1.0
556-61-6	Methyl isothiocyanate	1.0
	[Isothiocyanatomethane]	
75-86-5	2-Methyllactonitrile	1.0
80-62-6	Methyl methacrylate	1.0
924-42-5	N-Methylolacrylamide	1.0
298-00-0	Methyl parathion	1.0
109-06-8	2-Methylpyridine	1.0
872-50-4	N-Methyl-2-pyrrolidone	1.0
9006-42-2	Metiram	1.0
21087-64-9	Metribuzin	1.0
7786-34-7	Mevinphos	1.0
90-94-8	Michler's ketone	0.1
2212-67-1	Molinate	1.0
	(1H-Azepine-1-carbothioic acid, hexahydro-, S-ethyl ester)	
1313-27-5	Molybdenum trioxide	1.0
76-15-3	Monochloropentafluoroethane	1.0
	(CFC-115)	
150-68-5	Monuron	1.0
505-60-2	Mustard gas	0.1
	[Ethane, 1,1'-thiobis[2-chloro-]	
88671-89-0	Myclobutanil	1.0
	[.alpha.-Butyl-.alpha.-(4-chlorophenyl)-1H-1,2,4-triazole-1-propanenitrile]	
142-59-6	Nabam	1.0
300-76-5	Naled	1.0
91-20-3	Naphthalene	0.1
134-32-7	alpha-Naphthylamine	0.1
91-59-8	beta-Naphthylamine	0.1
7440-02-0	Nickel	0.1
1929-82-4	Nitrapyrin	1.0
	(2-Chloro-6-(trichloromethyl)pyridine)	

CAS Number	Chemical Name	De minimis % Limit
7697-37-2	Nitric acid	1.0
139-13-9	Nitrilotriacetic acid	0.1
100-01-6	p-Nitroaniline	1.0
91-23-6	o-Nitroanisole	0.1
99-59-2	5-Nitro-o-anisidine	1.0
98-95-3	Nitrobenzene	0.1
92-93-3	4-Nitrobiphenyl	0.1
1836-75-5	Nitrofen	0.1
	[Benzene, 2,4-dichloro-1-(4-nitrophenoxy)-]	
51-75-2	Nitrogen mustard	0.1
	[2-Chloro-N-(2-chloroethyl)-N-methylethanamine]	
55-63-0	Nitroglycerin	1.0
75-52-5	Nitromethane	0.1
88-75-5	2-Nitrophenol	1.0
100-02-7	4-Nitrophenol	1.0
79-46-9	2-Nitropropane	0.1
924-16-3	N-Nitrosodi-n-butylamine	0.1
55-18-5	N-Nitrosodiethylamine	0.1
62-75-9	N-Nitrosodimethylamine	0.1
86-30-6	N-Nitrosodiphenylamine	1.0
156-10-5	p-Nitrosodiphenylamine	1.0
621-64-7	N-Nitrosodi-n-propylamine	0.1
759-73-9	N-Nitroso-N-ethylurea	0.1
684-93-5	N-Nitroso-N-methylurea	0.1
4549-40-0	N-Nitrosomethylvinylamine	0.1
59-89-2	N-Nitrosomorpholine	0.1
16543-55-8	N-Nitrosonornicotine	0.1
100-75-4	N-Nitrosopiperidine	0.1
99-55-8	5-Nitro-o-toluidine	1.0
27314-13-2	Norflurazon	1.0
	[4-Chloro-5-(methylamino)-2-[3-(trifluoromethyl)phenyl]-3(2H)-pyridazinone]	
2234-13-1	Octachloronaphthalene	1.0
29082-74-4	Octachlorostyrene	*
19044-88-3	Oryzalin	1.0
	[4-(Dipropylamino)-3,5-dinitrobenzene sulfonamide]	
20816-12-0	Osmium tetroxide	1.0
301-12-2	Oxydemeton methyl	1.0
	[S-(2-(Ethylsulfinyl)ethyl) O,O-dimethyl ester phosphorothioic acid]	
19666-30-9	Oxydiazon	1.0
	[3-[2,4-Dichloro-5-(1-methylethoxy)phenyl]- 5-(1,1-dimethylethyl)-1,3,4-oxadiazol-2(3H)-one]	
42874-03-3	Oxyfluorfen	1.0
10028-15-6	Ozone	1.0
123-63-7	Paraldehyde	1.0
1910-42-5	Paraquat dichloride	1.0
56-38-2	Parathion	1.0
	[Phosphorothioic acid, O,O-diethyl-O-(4-nitrophenyl)ester]	
1114-71-2	Pebulate	1.0
	[Butylethylcarbamothioic acid S-propyl ester]	

Table II. EPCRA Section 313 Chemical List for Reporting Year 2012

CAS Number	Chemical Name	De minimis % Limit
40487-42-1	Pendimethalin [N-(1-Ethylpropyl)-3,4-dimethyl-2,6-dinitrobenzenamine]	*
608-93-5	Pentachlorobenzene	*
76-01-7	Pentachloroethane	1.0
87-86-5	Pentachlorophenol (PCP)	0.1
57-33-0	Pentobarbital sodium	1.0
79-21-0	Peracetic acid	1.0
594-42-3	Perchloromethyl mercaptan	1.0
52645-53-1	Permethrin [3-(2,2-Dichloroethenyl)-2,2-dimethylcyclopropanecarboxylic acid, (3-phenoxyphenyl) methyl ester]	1.0
85-01-8	Phenanthrene	1.0
108-95-2	Phenol	1.0
77-09-8	Phenolphthalein	0.1
26002-80-2	Phenothrin [2,2-Dimethyl-3-(2-methyl-1-propenyl)cyclopropanecarboxylic acid (3-phenoxyphenyl)methyl ester]	1.0
95-54-5	1,2-Phenylenediamine	1.0
108-45-2	1,3-Phenylenediamine	1.0
106-50-3	p-Phenylenediamine	1.0
615-28-1	1,2-Phenylenediamine dihydro-chloride	1.0
624-18-0	1,4-Phenylenediamine dihydro-chloride	1.0
90-43-7	2-Phenylphenol	1.0
57-41-0	Phenytoin	0.1
75-44-5	Phosgene	1.0
7803-51-2	Phosphine	1.0
7723-14-0	Phosphorus (yellow or white)	1.0
85-44-9	Phthalic anhydride	1.0
1918-02-1	Picloram	1.0
88-89-1	Picric acid	1.0
51-03-6	Piperonyl butoxide	1.0
29232-93-7	Pirimiphos methyl [O-(2-(Diethylamino)-6-methyl-4-pyrimidinyl)-O,O-dimethylphosphorothioate]	1.0
1336-36-3	Polychlorinated biphenyls (PCBs)	*
7758-01-2	Potassium bromate	0.1
128-03-0	Potassium dimethyldithio-carbamate	1.0
137-41-7	Potassium N-methyldithio-carbamate	1.0
41198-08-7	Profenofos [O-(4-Bromo-2-chlorophenyl)-O-ethyl-S-propyl phosphorothioate]	1.0
7287-19-6	Prometryn [N,N'-Bis(1-methylethyl)-6-methylthio-1,3,5-triazine-2,4-diamine]	1.0
23950-58-5	Pronamide	1.0
1918-16-7	Propachlor [2-Chloro-N-(1-methylethyl)-N-phenylacetamide]	1.0
1120-71-4	Propane sultone	0.1
709-98-8	Propanil [N-(3,4-Dichlorophenyl)propanamide]	1.0
2312-35-8	Propargite	1.0
107-19-7	Propargyl alcohol	1.0
31218-83-4	Propetamphos [3-[(Ethylamino)methoxyphosphinothioyl]oxy]-2-butenoic acid, 1-methylethyl ester]	1.0
60207-90-1	Propiconazole [1-[2-(2,4-Dichlorophenyl)-4-propyl-1,3-dioxolan-2-yl]-methyl-1H-1,2,4,-triazole]	1.0
57-57-8	beta-Propiolactone	0.1
123-38-6	Propionaldehyde	1.0
114-26-1	Propoxur [Phenol, 2-(1-methylethoxy)-, methylcarbamate]	1.0
115-07-1	Propylene (Propene)	1.0
75-55-8	Propyleneimine	0.1
75-56-9	Propylene oxide	0.1
110-86-1	Pyridine	1.0
91-22-5	Quinoline	1.0
106-51-4	Quinone	1.0
82-68-8	Quintozene (Pentachloronitrobenzene)	1.0
76578-14-8	Quizalofop-ethyl [2-[4-[(6-Chloro-2-quinoxalinyl)oxy]phenoxy] propanoic acid ethyl ester]	1.0
10453-86-8	Resmethrin [[5-(Phenylmethyl)-3-furanyl]methyl-2,2-dimethyl-3-(2-methyl-1-propenyl) cyclopropanecarboxylate]	1.0
81-07-2	Saccharin (only persons who manufacture are subject, no supplier notification)	1.0
94-59-7	Safrole	0.1
7782-49-2	Selenium	1.0
74051-80-2	Sethoxydim [2-[1-(Ethoxyimino)butyl]-5-[2-(ethylthio)propyl]-3-hydroxyl-2-cyclohexen-1-one]	1.0
7440-22-4	Silver	1.0
122-34-9	Simazine	1.0
26628-22-8	Sodium azide	1.0
1982-69-0	Sodium dicamba [3,6-Dichloro-2-methoxybenzoic acid, sodium salt]	1.0
128-04-1	Sodium dimethyldithiocarbamate	1.0
62-74-8	Sodium fluoroacetate	1.0
7632-00-0	Sodium nitrite	1.0
131-52-2	Sodium pentachlorophenate	1.0
132-27-4	Sodium o-phenylphenoxide	0.1
100-42-5	Styrene	0.1
96-09-3	Styrene oxide	0.1
7664-93-9	Sulfuric acid (acid aerosols including mists, vapors, gas, fog, and other airborne forms of any particle	1.0

Table II. EPCRA Section 313 Chemical List for Reporting Year 2012

CAS Number	Chemical Name	De minimis % Limit
	size)	
2699-79-8	Sulfuryl fluoride (Vikane)	1.0
35400-43-2	Sulprofos	1.0
	[O-Ethyl O-[4-(methylthio)phenyl] phosphorodithioic acid S-propylester]	
34014-18-1	Tebuthiuron	1.0
	[N-[5-(1,1-Dimethylethyl)-1,3,4-thiadiazol-2-yl]-N,N'-dimethylurea]	
3383-96-8	Temephos	1.0
5902-51-2	Terbacil	1.0
	[5-Chloro-3-(1,1-dimethylethyl)-6-methyl-2,4(1H,3H)-pyrimidinedione]	
79-94-7	Tetrabromobisphenol A	*
630-20-6	1,1,1,2-Tetrachloroethane	1.0
79-34-5	1,1,2,2-Tetrachloroethane	1.0
127-18-4	Tetrachloroethylene (Perchloroethylene)	0.1
354-11-0	1,1,1,2-Tetrachloro-2-fluoroethane (HCFC-121a)	1.0
354-14-3	1,1,2,2-Tetrachloro-1-fluoroethane (HCFC-121)	1.0
961-11-5	Tetrachlorvinphos	1.0
	[Phosphoric acid, 2-chloro-1-(2,4,5-trichlorophenyl) ethenyl dimethyl ester]	
64-75-5	Tetracycline hydrochloride	1.0
116-14-3	Tetrafluoroethylene	0.1
509-14-8	Tetranitromethane	0.1
7696-12-0	Tetramethrin	1.0
	[2,2-Dimethyl-3-(2-methyl-1-propenyl) cyclopropanecarboxylic acid (1,3,4,5,6,7-hexahydro-1,3-dioxo-2H-isoindol-2-yl)methyl ester]	
7440-28-0	Thallium	1.0
148-79-8	Thiabendazole	1.0
	[2-(4-Thiazolyl)-1H-benzimidazole]	
62-55-5	Thioacetamide	0.1
28249-77-6	Thiobencarb	1.0
	[Carbamic acid, diethylthio-, S-(p-chlorobenzyl)ester]	
139-65-1	4,4'-Thiodianiline	0.1
59669-26-0	Thiodicarb	1.0
23564-06-9	Thiophanate ethyl	1.0
	[[1,2-Phenylenebis(iminocarbonothioyl)] biscarbamic acid diethylester]	
23564-05-8	Thiophanate methyl	1.0
79-19-6	Thiosemicarbazide	1.0
62-56-6	Thiourea	0.1
137-26-8	Thiram	1.0
1314-20-1	Thorium dioxide	1.0
7550-45-0	Titanium tetrachloride	1.0
108-88-3	Toluene	1.0
584-84-9	Toluene-2,4-diisocyanate	0.1
91-08-7	Toluene-2,6-diisocyanate	0.1
26471-62-5	Toluene diisocyanate (mixed isomers)	0.1
95-53-4	o-Toluidine	0.1
636-21-5	o-Toluidine hydrochloride	0.1
8001-35-2	Toxaphene	*

CAS Number	Chemical Name	De minimis % Limit
43121-43-3	Triadimefon	1.0
	[1-(4-Chlorophenoxy)-3,3-di-methyl-1-(1H-1,2,4- triazol-1-yl)-2-butanone]	
2303-17-5	Triallate	1.0
68-76-8	Triaziquone	1.0
	[2,5-Cyclohexadiene-1,4-dione, 2,3,5-tris(1-aziridinyl)-]	
101200-48-0	Tribenuron methyl	1.0
	[2-[[[[(4-Methoxy-6-methyl-1,3,5-triazin-2-yl)-methylamino]-carbonyl]amino]sulfonyl] benzoic acid methyl ester)	
1983-10-4	Tributyltin fluoride	1.0
2155-70-6	Tributyltin methacrylate	1.0
78-48-8	S,S,S-Tributyltrithio-phosphate (DEF)	1.0
52-68-6	Trichlorfon	1.0
	[Phosphoric acid,(2,2,2-trichloro-l-hydroxy-ethyl)-, dimethyl ester]	
76-02-8	Trichloroacetyl chloride	1.0
120-82-1	1,2,4-Trichlorobenzene	1.0
71-55-6	1,1,1-Trichloroethane (Methyl chloroform)	1.0
79-00-5	1,1,2-Trichloroethane	1.0
79-01-6	Trichloroethylene	0.1
75-69-4	Trichlorofluoromethane (CFC-11)	1.0
95-95-4	2,4,5-Trichlorophenol	1.0
88-06-2	2,4,6-Trichlorophenol	0.1
96-18-4	1,2,3-Trichloropropane	0.1
57213-69-1	Triclopyr triethylammonium salt	1.0
121-44-8	Triethylamine	1.0
1582-09-8	Trifluralin	*
	[Benezeneamine, 2,6-dinitro-N,N-dipropyl-4-(trifluoromethyl)-]	
26644-46-2	Triforine	1.0
	[N,N'-[1,4-Piperazinediylbis-(2,2,2-trichloroethylidene)]bisformamide]	
95-63-6	1,2,4-Trimethylbenzene	1.0
2655-15-4	2,3,5-Trimethylphenyl methylcarbamate	1.0
639-58-7	Triphenyltin chloride	1.0
76-87-9	Triphenyltin hydroxide	1.0
126-72-7	Tris(2,3-dibromopropyl) phosphate	0.1
72-57-1	Trypan blue	0.1
51-79-6	Urethane (Ethyl carbamate)	0.1
7440-62-2	Vanadium (except when contained in an alloy)	1.0
50471-44-8	Vinclozolin	1.0
	[3-(3,5-Dichlorophenyl)-5-ethenyl-5-methyl-2,4-oxazolidinedione]	
108-05-4	Vinyl acetate	0.1
593-60-2	Vinyl bromide	0.1
75-01-4	Vinyl chloride	0.1
75-02-5	Vinyl fluoride	0.1
75-35-4	Vinylidene chloride	1.0
108-38-3	m-Xylene	1.0
95-47-6	o-Xylene	1.0
106-42-3	p-Xylene	1.0

Table II. EPCRA Section 313 Chemical List for Reporting Year 2012

CAS Number	Chemical Name	De minimis % Limit
1330-20-7	Xylene (mixed isomers)	1.0
87-62-7	2,6-Xylidine	0.1
7440-66-6	Zinc (fume or dust)	1.0
12122-67-7	Zineb [Carbamodithioic acid, 1,2-ethanediyibis-, zinc complex]	1.0

b. Individually Listed Toxic Chemicals Arranged by CAS Number

CAS Number	Chemical Name	De minimis % Limit
Arranged by CAS Number		
50-00-0	Formaldehyde	0.1
51-03-6	Piperonyl butoxide	1.0
51-21-8	Fluorouracil (5-Fluorouracil)	1.0
51-28-5	2,4-Dinitrophenol	1.0
51-75-2	Nitrogen mustard [2-Chloro-N-(2-chloroethyl)-N-methylethanamine]	0.1
51-79-6	Urethane (Ethyl carbamate)	0.1
52-68-6	Trichlorfon [Phosphonic acid, (2,2,2-trichloro-1-hydroxyethyl)-, dimethyl ester]	1.0
52-85-7	Famphur	1.0
53-96-3	2-Acetylaminofluorene	0.1
55-18-5	N-Nitrosodiethylamine	0.1
55-21-0	Benzamide	1.0
55-38-9	Fenthion [O,O-Dimethyl O-[3-methyl-4-(methylthio)phenyl] ester, phosphorothioic acid]	1.0
55-63-0	Nitroglycerin	1.0
56-23-5	Carbon tetrachloride	0.1
56-35-9	Bis(tributyltin) oxide	1.0
56-38-2	Parathion [Phosphorothioic acid, O,O-diethyl-O-(4-nitrophenyl) ester]	1.0
57-14-7	1,1-Dimethylhydrazine	0.1
57-33-0	Pentobarbital sodium	1.0
57-41-0	Phenytoin	0.1
57-57-8	beta-Propiolactone	0.1
57-74-9	Chlordane [4,7-Methanoindan, 1,2,4,5,6,7,8,8-octachloro-2,3,3a,4,7,7a-hexahydro-]	*
58-89-9	Lindane [Cyclohexane, 1,2,3,4,5,6-hexachloro-, (1.alpha.,2.alpha.,3.beta.,4.alpha,5.alpha.,6.beta.)-]	0.1
59-89-2	N-Nitrosomorpholine	0.1
60-09-3	4-Aminoazobenzene	0.1
60-11-7	4-Dimethylaminoazobenzene	0.1
60-34-4	Methyl hydrazine	1.0
60-35-5	Acetamide	0.1
60-51-5	Dimethoate	1.0
61-82-5	Amitrole	0.1
62-53-3	Aniline	1.0

CAS Number	Chemical Name	De minimis % Limit
Arranged by CAS Number		
62-55-5	Thioacetamide	0.1
62-56-6	Thiourea	0.1
62-73-7	Dichlorvos [Phosphoric acid, 2,2-dichloroethenyl dimethyl ester]	0.1
62-74-8	Sodium fluoroacetate	1.0
62-75-9	N-Nitrosodimethylamine	0.1
63-25-2	Carbaryl [1-Naphthalenol, methylcarbamate]	1.0
64-18-6	Formic acid	1.0
64-67-5	Diethyl sulfate	0.1
64-75-5	Tetracycline hydrochloride	1.0
67-56-1	Methanol	1.0
67-63-0	Isopropyl alcohol (only persons who manufacture by the strong acid process are subject, no supplier notification)	1.0
67-66-3	Chloroform	0.1
67-72-1	Hexachloroethane	0.1
68-12-2	N,N-Dimethylformamide	1.0
68-76-8	Triaziquone [2,5-Cyclohexadiene-1,4-dione, 2,3,5-tris(1-aziridinyl)-]	1.0
70-30-4	Hexachlorophene	1.0
71-36-3	n-Butyl alcohol	1.0
71-43-2	Benzene	0.1
71-55-6	1,1,1-Trichloroethane (Methyl chloroform)	1.0
72-43-5	Methoxychlor [Benzene, 1,1'-(2,2,2-trichloroethylidene)bis[4-methoxy-]	*
72-57-1	Trypan blue	0.1
74-83-9	Bromomethane (Methyl bromide)	1.0
74-85-1	Ethylene	1.0
74-87-3	Chloromethane (Methyl chloride)	1.0
74-88-4	Methyl iodide	1.0
74-90-8	Hydrogen cyanide	1.0
74-95-3	Methylene bromide	1.0
75-00-3	Chloroethane (Ethyl chloride)	1.0
75-01-4	Vinyl chloride	0.1
75-02-5	Vinyl fluoride	0.1
75-05-8	Acetonitrile	1.0
75-07-0	Acetaldehyde	0.1
75-09-2	Dichloromethane (Methylene chloride)	0.1
75-15-0	Carbon disulfide	1.0
75-21-8	Ethylene oxide	0.1
75-25-2	Bromoform (Tribromomethane)	1.0
75-27-4	Dichlorobromomethane	0.1
75-34-3	Ethylidene dichloride	1.0
75-35-4	Vinylidene chloride	1.0
75-43-4	Dichlorofluoromethane (HCFC-21)	1.0
75-44-5	Phosgene	1.0
75-45-6	Chlorodifluoromethane (HCFC-22)	1.0
75-52-5	Nitromethane	0.1

Table II. EPCRA Section 313 Chemical List for Reporting Year 2012

CAS Number	Chemical Name	De minimis % Limit	CAS Number	Chemical Name	De minimis % Limit
	Arranged by CAS Number			*Arranged by CAS Number*	
75-55-8	Propyleneimine	0.1	81-49-2	1-Amino-2,4-dibromoanthraquinone	0.1
75-56-9	Propylene oxide	0.1	81-88-9	C.I. Food Red 15	1.0
75-63-8	Bromotrifluoromethane (Halon 1301)	1.0	82-28-0	1-Amino-2-methylanthraquinone	0.1
75-65-0	tert-Butyl alcohol	1.0	82-68-8	Quintozene [Pentachloronitrobenzene]	1.0
75-68-3	1-Chloro-1,1-difluoroethane (HCFC-142b)	1.0	84-74-2	Dibutyl phthalate	1.0
75-69-4	Trichlorofluoromethane (CFC-11)	1.0	85-01-8	Phenanthrene	1.0
75-71-8	Dichlorodifluoromethane (CFC-12)	1.0	85-44-9	Phthalic anhydride	1.0
75-72-9	Chlorotrifluoromethane (CFC-13)	1.0	86-30-6	N-Nitrosodiphenylamine	1.0
75-86-5	2-Methyllactonitrile	1.0	87-62-7	2,6-Xylidine	0.1
75-88-7	2-Chloro-1,1,1-trifluoroethane (HCFC-133a)	1.0	87-68-3	Hexachloro-1,3-butadiene	1.0
76-01-7	Pentachloroethane	1.0	87-86-5	Pentachlorophenol (PCP)	0.1
76-02-8	Trichloroacetyl chloride	1.0	88-06-2	2,4,6-Trichlorophenol	0.1
76-06-2	Chloropicrin	1.0	88-75-5	2-Nitrophenol	1.0
76-13-1	Freon 113 [Ethane, 1,1,2-trichloro-1,2,2,-trifluoro-]	1.0	88-85-7	Dinitrobutyl phenol (Dinoseb)	1.0
76-14-2	Dichlorotetrafluoroethane (CFC-114)	1.0	88-89-1	Picric acid	1.0
76-15-3	Monochloropentafluoroethane (CFC-115)	1.0	90-04-0	o-Anisidine	0.1
76-44-8	Heptachlor [1,4,5,6,7,8,8-Heptachloro-3a,4,7,7a-tetrahydro-4,7-methano-1H-indene]	*	90-43-7	2-Phenylphenol	1.0
			90-94-8	Michler's ketone	0.1
			91-08-7	Toluene-2,6-diisocyanate	0.1
76-87-9	Triphenyltin hydroxide	1.0	91-20-3	Naphthalene	0.1
77-09-8	Phenolphthalein	0.1	91-22-5	Quinoline	1.0
77-47-4	Hexachlorocyclopentadiene	1.0	91-23-6	o-Nitroanisole	0.1
77-73-6	Dicyclopentadiene	1.0	91-59-8	beta-Naphthylamine	0.1
77-78-1	Dimethyl sulfate	0.1	91-94-1	3,3'-Dichlorobenzidine	0.1
78-48-8	S,S,S-Tributyltrithiophosphate (DEF)	1.0	92-52-4	Biphenyl	1.0
			92-67-1	4-Aminobiphenyl	0.1
78-79-5	Isoprene	0.1	92-87-5	Benzidine	0.1
78-84-2	Isobutyraldehyde	1.0	92-93-3	4-Nitrobiphenyl	0.1
78-87-5	1,2-Dichloropropane	1.0	93-15-2	Methyleugenol	0.1
78-88-6	2,3-Dichloropropene	1.0	93-65-2	Mecoprop	0.1
78-92-2	sec-Butyl alcohol	1.0	94-11-1	2,4-D isopropyl ester	0.1
79-00-5	1,1,2-Trichloroethane	1.0	94-36-0	Benzoyl peroxide	1.0
79-01-6	Trichloroethylene	0.1	94-58-6	Dihydrosafrole	0.1
79-06-1	Acrylamide	0.1	94-59-7	Safrole	0.1
79-10-7	Acrylic acid	1.0	94-74-6	Methoxone ((4-Chloro-2-methylphenoxy) acetic acid) (MCPA)	0.1
79-11-8	Chloroacetic acid	1.0			
79-19-6	Thiosemicarbazide	1.0			
79-21-0	Peracetic acid	1.0	94-75-7	2,4-D [Acetic acid, (2,4-dichlorophenoxy)-]	0.1
79-22-1	Methyl chlorocarbonate	1.0			
79-34-5	1,1,2,2-Tetrachloroethane	1.0	94-80-4	2,4-D butyl ester	0.1
79-44-7	Dimethylcarbamyl chloride	0.1	94-82-6	2,4-DB	1.0
79-46-9	2-Nitropropane	0.1	95-47-6	o-Xylene	1.0
79-94-7	Tetrabromobisphenol A	*	95-48-7	o-Cresol	1.0
80-05-7	4,4'-Isopropylidenediphenol	1.0	95-50-1	1,2-Dichlorobenzene	1.0
80-15-9	Cumene hydroperoxide	1.0	95-53-4	o-Toluidine	0.1
80-62-6	Methyl methacrylate	1.0	95-54-5	1,2-Phenylenediamine	1.0
			95-63-6	1,2,4-Trimethylbenzene	1.0
			95-69-2	p-Chloro-o-toluidine	0.1
			95-80-7	2,4-Diaminotoluene	0.1
81-07-2	Saccharin (only persons who manufacture are subject, no supplier notification)	1.0	95-95-4	2,4,5-Trichlorophenol	1.0
			96-09-3	Styrene oxide	0.1
			96-12-8	1,2-Dibromo-3-chloropropane (DBCP)	0.1
			96-18-4	1,2,3-Trichloropropane	0.1
			96-33-3	Methyl acrylate	1.0
			96-45-7	Ethylene thiourea	0.1

Table II. EPCRA Section 313 Chemical List for Reporting Year 2012

CAS Number	Chemical Name	De minimis % Limit
	Arranged by CAS Number	
97-23-4	Dichlorophene [2,2'-Methylenebis(4-chlorophenol)]	1.0
97-56-3	C.I. Solvent Yellow 3	0.1
98-07-7	Benzoic trichloride (Benzotrichloride)	0.1
98-82-8	Cumene	1.0
98-86-2	Acetophenone	1.0
98-87-3	Benzal chloride	1.0
98-88-4	Benzoyl chloride	1.0
98-95-3	Nitrobenzene	0.1
99-30-9	Dichloran [2,6-Dichloro-4-nitroaniline]	1.0
99-55-8	5-Nitro-o-toluidine	1.0
99-59-2	5-Nitro-o-anisidine	1.0
99-65-0	m-Dinitrobenzene	1.0
100-01-6	p-Nitroaniline	1.0
100-02-7	4-Nitrophenol	1.0
100-25-4	p-Dinitrobenzene	1.0
100-41-4	Ethylbenzene	0.1
100-42-5	Styrene	0.1
100-44-7	Benzyl chloride	1.0
100-75-4	N-Nitrosopiperidine	0.1
101-05-3	Anilazine [4,6-Dichloro-N-(2-chlorophenyl)-1,3,5-triazin-2-amine]	1.0
101-14-4	4,4'-Methylenebis(2-chloroaniline) (MBOCA)	0.1
101-61-1	4,4'-Methylenebis(N,N-dimethyl)benzenamine	0.1
101-77-9	4,4'-Methylenedianiline	0.1
101-80-4	4,4'-Diaminodiphenyl ether	0.1
101-90-6	Diglycidyl resorcinol ether	0.1
104-12-1	p-Chlorophenyl isocyanate	1.0
104-94-9	p-Anisidine	1.0
105-67-9	2,4-Dimethylphenol	1.0
106-42-3	p-Xylene	1.0
106-44-5	p-Cresol	1.0
106-46-7	1,4-Dichlorobenzene	0.1
106-47-8	p-Chloroaniline	0.1
106-50-3	p-Phenylenediamine	1.0
106-51-4	Quinone	1.0
106-88-7	1,2-Butylene oxide	0.1
106-89-8	Epichlorohydrin	0.1
106-93-4	1,2-Dibromoethane (Ethylene dibromide)	0.1
106-99-0	1,3-Butadiene	0.1
107-02-8	Acrolein	1.0
107-05-1	Allyl chloride	1.0
107-06-2	1,2-Dichloroethane (Ethylene dichloride)	0.1
107-11-9	Allylamine	1.0
107-13-1	Acrylonitrile	0.1
107-18-6	Allyl alcohol	1.0
107-19-7	Propargyl alcohol	1.0
107-21-1	Ethylene glycol	1.0
107-30-2	Chloromethyl methyl ether	0.1
108-05-4	Vinyl acetate	0.1

CAS Number	Chemical Name	De minimis % Limit
	Arranged by CAS Number	
108-10-1	Methyl isobutyl ketone	1.0
108-31-6	Maleic anhydride	1.0
108-38-3	m-Xylene	1.0
108-39-4	m-Cresol	1.0
108-45-2	1,3-Phenylenediamine	1.0
108-60-1	Bis(2-chloro-1-methylethyl) ether	1.0
108-88-3	Toluene	1.0
108-90-7	Chlorobenzene	1.0
108-93-0	Cyclohexanol	1.0
108-95-2	Phenol	1.0
109-06-8	2-Methylpyridine	1.0
109-77-3	Malononitrile	1.0
109-86-4	2-Methoxyethanol	1.0
110-00-9	Furan	0.1
110-54-3	n-Hexane	1.0
110-57-6	trans-1,4-Dichloro-2-butene	1.0
110-80-5	2-Ethoxyethanol	1.0
110-82-7	Cyclohexane	1.0
110-86-1	Pyridine	1.0
111-42-2	Diethanolamine	1.0
111-44-4	Bis(2-chloroethyl) ether	1.0
111-91-1	Bis(2-chloroethoxy) methane	1.0
114-26-1	Propoxur [Phenol, 2-(1-methylethoxy)-, methylcarbamate]	1.0
115-07-1	Propylene (Propene)	1.0
115-28-6	Chlorendic acid	0.1
115-32-2	Dicofol [Benzenemethanol, 4-chloro-.alpha.-4-(chlorophenyl)-.alpha.-(trichloromethyl)-]	1.0
116-06-3	Aldicarb	1.0
116-14-3	Tetrafluoroethylene	0.1
117-79-3	2-Aminoanthraquinone	0.1
117-81-7	Di(2-ethylhexyl) phthalate	0.1
118-74-1	Hexachlorobenzene	*
119-90-4	3,3'-Dimethoxybenzidine	0.1
119-93-7	3,3'-Dimethylbenzidine (o-Tolidine)	0.1
120-12-7	Anthracene	1.0
120-36-5	2,4-DP	0.1
120-58-1	Isosafrole	1.0
120-71-8	p-Cresidine	0.1
120-80-9	Catechol	0.1
120-82-1	1,2,4-Trichlorobenzene	1.0
120-83-2	2,4-Dichlorophenol	1.0
121-14-2	2,4-Dinitrotoluene	0.1
121-44-8	Triethylamine	1.0
121-69-7	N,N-Dimethylaniline	1.0
121-75-5	Malathion	1.0
122-34-9	Simazine	1.0
122-39-4	Diphenylamine	1.0
122-66-7	1,2-Diphenylhydrazine (Hydrazobenzene)	0.1
123-31-9	Hydroquinone	1.0
123-38-6	Propionaldehyde	1.0
123-63-7	Paraldehyde	1.0
123-72-8	Butyraldehyde	1.0

Table II. EPCRA Section 313 Chemical List for Reporting Year 2012

CAS Number	Chemical Name	De minimis % Limit
	Arranged by CAS Number	
123-91-1	1,4-Dioxane	0.1
124-40-3	Dimethylamine	1.0
124-73-2	Dibromotetrafluoroethane (Halon 2402)	1.0
126-72-7	Tris(2,3-dibromopropyl) phosphate	0.1
126-98-7	Methacrylonitrile	1.0
126-99-8	Chloroprene	0.1
127-18-4	Tetrachloroethylene (Perchloroethylene)	0.1
128-03-0	Potassium dimethyldithiocarbamate	1.0
128-04-1	Sodium dimethyldithiocarbamate	1.0
128-66-5	C.I. Vat Yellow 4	1.0
131-11-3	Dimethyl phthalate	1.0
131-52-2	Sodium pentachlorophenate	1.0
132-27-4	Sodium o-phenylphenoxide	0.1
132-64-9	Dibenzofuran	1.0
133-06-2	Captan [1H-Isoindole-1,3(2H)-dione, 3a,4,7,7a-tetrahydro-2-[(trichloromethyl)thio]-]	1.0
133-07-3	Folpet	1.0
133-90-4	Chloramben [Benzoic acid, 3-amino-2,5-dichloro-]	1.0
134-29-2	o-Anisidine hydrochloride	0.1
134-32-7	alpha-Naphthylamine	0.1
135-20-6	Cupferron [Benzeneamine, N-hydroxy-N-nitroso, ammonium salt]	0.1
136-45-8	Dipropyl isocinchomeronate	1.0
137-26-8	Thiram	1.0
137-41-7	Potassium N-methyldithio-carbamate	1.0
137-42-8	Metham sodium (Sodium methyldithiocarbamate)	1.0
138-93-2	Disodium cyanodithioimido-carbonate	1.0
139-13-9	Nitrilotriacetic acid	0.1
139-65-1	4,4'-Thiodianiline	0.1
140-88-5	Ethyl acrylate	0.1
141-32-2	Butyl acrylate	1.0
142-59-6	Nabam	1.0
148-79-8	Thiabendazole [2-(4-Thiazolyl)-1H-benzimidazole]	1.0
149-30-4	2-Mercaptobenzothiazole (MBT)	1.0
150-50-5	Merphos	1.0
150-68-5	Monuron	1.0
151-56-4	Ethyleneimine (Aziridine)	0.1
156-10-5	p-Nitrosodiphenylamine	1.0
156-62-7	Calcium cyanamide	1.0
191-24-2	Benzo(g,h,i)perylene	*
298-00-0	Methyl parathion	1.0
300-76-5	Naled	1.0
301-12-2	Oxydemeton methyl [S-(2-(Ethylsulfinyl)ethyl) O,O-dimethyl ester phosphorothioic acid]	1.0

CAS Number	Chemical Name	De minimis % Limit
	Arranged by CAS Number	
302-01-2	Hydrazine	0.1
306-83-2	2,2-Dichloro-1,1,1-trifluoroethane (HCFC-123)	1.0
309-00-2	Aldrin [1,4:5,8-Dimethanonaphthalene, 1,2,3,4,10,10-hexachloro-1,4,4a,5,8,8a-hexahydro-(1.alpha.,4.alpha.,4a.beta., 5.alpha.,8.alpha.,8a.beta.)-]	*
314-40-9	Bromacil (5-Bromo-6-methyl-3-(1-methylpropyl)-2,4(1H,3H)-pyrimidinedione)	1.0
319-84-6	alpha-Hexachlorocyclohexane	0.1
330-54-1	Diuron	1.0
330-55-2	Linuron	1.0
333-41-5	Diazinon	1.0
334-88-3	Diazomethane	1.0
353-59-3	Bromochlorodifluoromethane (Halon 1211)	1.0
354-11-0	1,1,1,2-Tetrachloro-2-fluoroethane (HCFC-121a)	1.0
354-14-3	1,1,2,2-Tetrachloro-1-fluoroethane (HCFC-121)	1.0
354-23-4	1,2-Dichloro-1,1,2-trifluoroethane (HCFC-123a)	1.0
354-25-6	1-Chloro-1,1,2,2-tetrafluoroethane (HCFC-124a)	1.0
357-57-3	Brucine	1.0
422-44-6	1,2-Dichloro-1,1,2,3,3-pentafluoropropane (HCFC-225bb)	1.0
422-48-0	2,3-Dichloro-1,1,1,2,3-pentafluoropropane (HCFC-225ba)	1.0
422-56-0	3,3-Dichloro-1,1,1,2,2-pentafluoropropane (HCFC-225ca)	1.0
431-86-7	1,2-Dichloro-1,1,3,3,3-pentafluoropropane (HCFC-225da)	1.0
460-35-5	3-Chloro-1,1,1-trifluoropropane (HCFC-253fb)	1.0
463-58-1	Carbonyl sulfide	1.0
465-73-6	Isodrin	*
492-80-8	C.I. Solvent Yellow 34 (Auramine)	0.1
505-60-2	Mustard gas [Ethane, 1,1'-thiobis[2-chloro-]	0.1
507-55-1	1,3-Dichloro-1,1,2,2,3-pentafluoropropane (HCFC-225cb)	1.0
509-14-8	Tetranitromethane	0.1
510-15-6	Chlorobenzilate [Benzeneacetic acid, 4-chloro-.alpha.-(4-chlorophenyl)-.alpha.-hydroxy-, ethyl ester]	1.0
528-29-0	o-Dinitrobenzene	1.0
532-27-4	2-Chloroacetophenone	1.0
533-74-4	Dazomet (Tetrahydro-3,5-dimethyl-2H-1,3,5-thiadiazine-2-thione)	1.0

Table II. EPCRA Section 313 Chemical List for Reporting Year 2012

CAS Number	Chemical Name	De minimis % Limit
	Arranged by CAS Number	
534-52-1	4,6-Dinitro-o-cresol	1.0
540-59-0	1,2-Dichloroethylene	1.0
541-41-3	Ethyl chloroformate	1.0
541-53-7	2,4-Dithiobiuret	1.0
541-73-1	1,3-Dichlorobenzene	1.0
542-75-6	1,3-Dichloropropylene	0.1
542-76-7	3-Chloropropionitrile	1.0
542-88-1	Bis(chloromethyl) ether	0.1
554-13-2	Lithium carbonate	1.0
556-52-5	Glycidol	0.1
556-61-6	Methyl isothiocyanate [Isothiocyanatomethane]	1.0
563-47-3	3-Chloro-2-methyl-1-propene	0.1
569-64-2	C.I. Basic Green 4	1.0
584-84-9	Toluene-2,4-diisocyanate	0.1
593-60-2	Vinyl bromide	0.1
594-42-3	Perchloromethyl mercaptan	1.0
606-20-2	2,6-Dinitrotoluene	0.1
608-93-5	Pentachlorobenzene	*
612-82-8	3,3'-Dimethylbenzidine dihydrochloride (o-Tolidine dihydrochloride)	0.1
612-83-9	3,3'-Dichlorobenzidine dihydrochloride	0.1
615-05-4	2,4-Diaminoanisole	0.1
615-28-1	1,2-Phenylenediamine dihydrochloride	1.0
621-64-7	N-Nitrosodi-n-propylamine	0.1
624-18-0	1,4-Phenylenediamine dihydrochloride	1.0
624-83-9	Methyl isocyanate	1.0
630-20-6	1,1,1,2-Tetrachloroethane	1.0
636-21-5	o-Toluidine hydrochloride	0.1
639-58-7	Triphenyltin chloride	1.0
680-31-9	Hexamethylphosphoramide	0.1
684-93-5	N-Nitroso-N-methylurea	0.1
709-98-8	Propanil (N-(3,4-Dichlorophenyl) propanamide)	1.0
759-73-9	N-Nitroso-N-ethylurea	0.1
759-94-4	Ethyl dipropylthiocarbamate (EPTC)	1.0
764-41-0	1,4-Dichloro-2-butene	1.0
812-04-4	1,1-Dichloro-1,2,2-trifluoroethane (HCFC-123b)	1.0
834-12-8	Ametryn (N-Ethyl-N'-(1-methylethyl)-6-(methylthio)-1,3,5,-triazine-2,4-diamine)	1.0
842-07-9	C.I. Solvent Yellow 14	1.0
872-50-4	N-Methyl-2-pyrrolidone	1.0
924-16-3	N-Nitrosodi-n-butylamine	0.1
924-42-5	N-Methylolacrylamide	1.0
957-51-7	Diphenamid	1.0
961-11-5	Tetrachlorvinphos [Phosphoric acid, 2-chloro-1-(2,4,5-trichlorophenyl)ethenyldimethyl ester]	1.0

CAS Number	Chemical Name	De minimis % Limit
	Arranged by CAS Number	
989-38-8	C.I. Basic Red 1	1.0
1114-71-2	Pebulate [Butylethylcarbamothioic acid S-propyl ester]	1.0
1120-71-4	Propane sultone	0.1
1134-23-2	Cycloate	1.0
1163-19-5	Decabromodiphenyl oxide	1.0
1313-27-5	Molybdenum trioxide	1.0
1314-20-1	Thorium dioxide	1.0
1319-77-3	Cresol (mixed isomers)	1.0
1320-18-9	2,4-D propylene glycol butyl ether ester	0.1
1330-20-7	Xylene (mixed isomers)	1.0
1332-21-4	Asbestos (friable)	0.1
1335-87-1	Hexachloronaphthalene	1.0
1336-36-3	Polychlorinated biphenyls (PCBs)	*
1344-28-1	Aluminum oxide (fibrous forms)	1.0
1464-53-5	Diepoxybutane	0.1
1563-66-2	Carbofuran	1.0
1582-09-8	Trifluralin [Benzeneamine, 2,6-dinitro-N,N-dipropyl-4-(trifluoromethyl)-]	*
1634-04-4	Methyl tert-butyl ether	1.0
1649-08-7	1,2-Dichloro-1,1-difluoroethane (HCFC-132b)	1.0
1689-84-5	Bromoxynil (3,5-Dibromo-4-hydroxybenzonitrile)	1.0
1689-99-2	Bromoxynil octanoate (Octanoic acid, 2,6-dibromo-4-cyanophenyl ester)	1.0
1717-00-6	1,1-Dichloro-1-fluoroethane (HCFC-141b)	1.0
1836-75-5	Nitrofen [Benzene, 2,4-dichloro-1-(4-nitrophenoxy)-]	0.1
1861-40-1	Benfluralin (N-Butyl-N-ethyl-2,6-dinitro-4-(trifluoromethyl)benzenamine)	1.0
1897-45-6	Chlorothalonil [1,3-Benzenedicarbonitrile, 2,4,5,6-tetrachloro-]	0.1
1910-42-5	Paraquat dichloride	1.0
1912-24-9	Atrazine (6-Chloro-N-ethyl-N'-(1-methylethyl)-1,3,5-triazine-2,4-diamine)	1.0
1918-00-9	Dicamba (3,6-Dichloro-2-methoxybenzoic acid)	1.0
1918-02-1	Picloram	1.0
1918-16-7	Propachlor [2-Chloro-N-(1-methylethyl)-N-phenylacetamide]	1.0
1928-43-4	2,4-D 2-ethylhexyl ester	0.1
1929-73-3	2,4-D butoxyethyl ester	0.1
1929-82-4	Nitrapyrin (2-Chloro-6-(trichloromethyl)pyridine)	1.0
1937-37-7	C.I. Direct Black 38	0.1

Table II. EPCRA Section 313 Chemical List for Reporting Year 2012

CAS Number	Chemical Name	*De minimis* % Limit
	Arranged by CAS Number	
1982-69-0	Sodium dicamba [3,6-Dichloro-2-methoxybenzoic acid, sodium salt]	1.0
1983-10-4	Tributyltin fluoride	1.0
2032-65-7	Methiocarb	1.0
2155-70-6	Tributyltin methacrylate	1.0
2164-07-0	Dipotassium endothall [7-Oxabicyclo(2.2.1)heptane-2,3-dicarboxylic acid, dipotassium salt]	1.0
2164-17-2	Fluometuron [Urea, N,N-dimethyl-N'-[3-(trifluoromethyl)phenyl]-]	1.0
2212-67-1	Molinate (1H-Azepine-1-carbothioic acid, hexahydro-S-ethyl ester)	1.0
2234-13-1	Octachloronaphthalene	1.0
2300-66-5	Dimethylamine dicamba	1.0
2303-16-4	Diallate [Carbamothioic acid, bis(1-methylethyl)-S-(2,3-dichloro-2-propenyl) ester]	1.0
2303-17-5	Triallate	1.0
2312-35-8	Propargite	1.0
2439-01-2	Chinomethionat [6-Methyl-1,3-dithiolo[4,5-b]quinoxalin-2-one]	1.0
2439-10-3	Dodine [Dodecylguanidine monoacetate]	1.0
2524-03-0	Dimethyl chlorothiophosphate	1.0
2602-46-2	C.I. Direct Blue 6	0.1
2655-15-4	2,3,5-Trimethylphenyl methyl carbamate	1.0
2699-79-8	Sulfuryl fluoride (Vikane)	1.0
2702-72-9	2,4-D sodium salt	0.1
2832-40-8	C.I. Disperse Yellow 3	1.0
2837-89-0	2-Chloro-1,1,1,2-tetrafluoroethane (HCFC-124)	1.0
2971-38-2	2,4-D Chlorocrotyl ester	0.1
3118-97-6	C.I. Solvent Orange 7	1.0
3296-90-0	2,2-bis(Bromomethyl)-1,3-propanediol	0.1
3383-96-8	Temephos	1.0
3653-48-3	Methoxone sodium salt ((4-Chloro-2-methylphenoxy) acetate sodium salt)	0.1
3761-53-3	C.I. Food Red 5	0.1
4080-31-3	1-(3-Chloroallyl)-3,5,7-triaza-1-azoniaadamantane chloride	1.0
4170-30-3	Crotonaldehyde	1.0
4549-40-0	N-Nitrosomethylvinylamine	0.1
4680-78-8	C.I. Acid Green 3	1.0
5234-68-4	Carboxin (5,6-Dihydro-2-methyl-N-phenyl-1,4-oxathiin-3-carboxamide)	1.0
5598-13-0	Chlorpyrifos methyl [O,O-Dimethyl-O-(3,5,6-trichloro-2-pyridyl)phosphorothioate]	1.0

CAS Number	Chemical Name	*De minimis* % Limit
	Arranged by CAS Number	
5902-51-2	Terbacil [5-Chloro-3-(1,1-dimethylethyl)-6-methyl-2,4(1H,3H)-pyrimidinedione]	1.0
6459-94-5	C.I. Acid Red 114	0.1
7287-19-6	Prometryn [N,N'-Bis(1-methylethyl)-6-methylthio-1,3,5-triazine-2,4-diamine]	1.0
7429-90-5	Aluminum (fume or dust)	1.0
7439-92-1	Lead (when lead is contained in stainless steel, brass or bronze alloys the *de minimis* level is 0.1)	*
7439-96-5	Manganese	1.0
7439-97-6	Mercury	*
7440-02-0	Nickel	0.1
7440-22-4	Silver	1.0
7440-28-0	Thallium	1.0
7440-36-0	Antimony	1.0
7440-38-2	Arsenic	0.1
7440-39-3	Barium	1.0
7440-41-7	Beryllium	0.1
7440-43-9	Cadmium	0.1
7440-47-3	Chromium	1.0
7440-48-4	Cobalt	0.1
7440-50-8	Copper	1.0
7440-62-2	Vanadium (except when contained in an alloy)	1.0
7440-66-6	Zinc (fume or dust)	1.0
7550-45-0	Titanium tetrachloride	1.0
7632-00-0	Sodium nitrite	1.0
7637-07-2	Boron trifluoride	1.0
7647-01-0	Hydrochloric acid (acid aerosols including mists, vapors, gas, fog, and other airborne forms of any particle size)	1.0
7664-39-3	Hydrogen fluoride	1.0
7664-41-7	Ammonia (includes anhydrous ammonia and aqueous ammonia from water dissociable ammonium salts and other sources; 10 percent of total aqueous ammonia is reportable under this listing)	1.0
7664-93-9	Sulfuric acid (acid aerosols including mists, vapors, gas, fog, and other airborne forms of any particle size)	1.0
7696-12-0	Tetramethrin [2,2-Dimethyl-3-(2-methyl-1-propenyl)cyclopropanecarboxylic acid (1,3,4,5,6,7-hexahydro-1,3-dioxo-2H-isoindol-2-yl)methyl ester]	1.0
7697-37-2	Nitric acid	1.0
7723-14-0	Phosphorus (yellow or white)	1.0
7726-95-6	Bromine	1.0
7758-01-2	Potassium bromate	0.1
7782-41-4	Fluorine	1.0
7782-49-2	Selenium	1.0

Table II. EPCRA Section 313 Chemical List for Reporting Year 2012

CAS Number	Chemical Name	De minimis % Limit
	Arranged by CAS Number	
7782-50-5	Chlorine	1.0
7783-06-4	Hydrogen sulfide	1.0
7786-34-7	Mevinphos	1.0
7803-51-2	Phosphine	1.0
8001-35-2	Toxaphene	*
8001-58-9	Creosote	0.1
9006-42-2	Metiram	1.0
10028-15-6	Ozone	1.0
10034-93-2	Hydrazine sulfate	0.1
10049-04-4	Chlorine dioxide	1.0
10061-02-6	trans-1,3-Dichloropropene	0.1
10294-34-5	Boron trichloride	1.0
10453-86-8	Resmethrin [[5-(Phenylmethyl)-3-furanyl]methyl- 2,2-dimethyl-3-(2-methyl-1-propenyl) cyclopropanecarboxylate]]	1.0
12122-67-7	Zineb [Carbamodithioic acid, 1,2-ethanediylbis-, zinc complex]	1.0
12427-38-2	Maneb [Carbamodithioic acid, 1,2-ethanediylbis-, manganese complex]	1.0
13194-48-4	Ethoprop [Phosphorodithioic acid O-ethyl S,S-dipropyl ester]	1.0
13356-08-6	Fenbutatin oxide (Hexakis(2-methyl-2-phenylpropyl) distannoxane)	1.0
13463-40-6	Iron pentacarbonyl	1.0
13474-88-9	1,1-Dichloro-1,2,2,3,3-pentafluoropropane (HCFC-225cc)	1.0
13684-56-5	Desmedipham	1.0
14484-64-1	Ferbam [Tris(dimethylcarbamodithioato-☐,☐' ▯▯▯▯▯	1.0
15972-60-8	Alachlor	1.0
16071-86-6	C.I. Direct Brown 95	0.1
16543-55-8	N-Nitrosonornicotine	0.1
17804-35-2	Benomyl	1.0
19044-88-3	Oryzalin [4-(Dipropylamino)-3,5-dinitrobenzenesulfonamide]	1.0
19666-30-9	Oxydiazon [3-[2,4-Dichloro-5-(1-methylethoxy) phenyl]-5-(1,1-dimethylethyl)-1,3,4-oxadiazol-2(3H)-one]	1.0
20325-40-0	3,3'-Dimethoxybenzidine dihydrochloride (o-Dianisidine dihydrochloride)	0.1
20354-26-1	Methazole [2-(3,4-Dichlorophenyl)-4-methyl-1,2,4-oxadiazolidine-3,5-dione]	1.0
20816-12-0	Osmium tetroxide	1.0
20859-73-8	Aluminum phosphide	1.0
21087-64-9	Metribuzin	1.0
21725-46-2	Cyanazine	1.0
22781-23-3	Bendiocarb [2,2-Dimethyl-1,3-benzodioxol-4-ol	1.0

CAS Number	Chemical Name	De minimis % Limit
	Arranged by CAS Number	
	methylcarbamate]	
23564-05-8	Thiophanate methyl	1.0
23564-06-9	Thiophanate ethyl [[1,2-Phenylenebis(iminocarbonothioyl)] biscarbamic acid diethyl ester]	1.0
23950-58-5	Pronamide	1.0
25311-71-1	Isofenphos [2-[[Ethoxyl[(1-methylethyl)-amino]phosphinothioyl]oxy]benzoic acid 1-methylethyl ester]	1.0
25321-14-6	Dinitrotoluene (mixed isomers)	1.0
25321-22-6	Dichlorobenzene (mixed isomers)	0.1
25376-45-8	Diaminotoluene (mixed isomers)	0.1
26002-80-2	Phenothrin [2,2-Dimethyl-3-(2-methyl-1-propenyl)cyclopropanecarboxylic acid (3-phenoxyphenyl)methyl ester]	1.0
26471-62-5	Toluene diisocyanate (mixed isomers)	0.1
26628-22-8	Sodium azide	1.0
26644-46-2	Triforine [N,N'-[1,4-Piperazinediylbis (2,2,2-trichloroethylidene)]bisformamide]	1.0
27314-13-2	Norflurazon [4-Chloro-5-(methylamino)-2-[3-(trifluoromethyl)phenyl]-3(2H)-pyridazinone]	1.0
28057-48-9	d-trans-Allethrin [d-trans-Chrysanthemic acid of d-allethrone]	1.0
28249-77-6	Thiobencarb [Carbamic acid, diethylthio-, S-(p-chlorobenzyl)ester]	1.0
28407-37-6	C.I. Direct Blue 218	1.0
29082-74-4	Octachlorostyrene	*
29232-93-7	Pirimiphos methyl [O-(2-(Diethylamino)-6-methyl-4-pyrimidinyl)-O,O-dimethylphosphorothioate]	1.0
30560-19-1	Acephate (Acetylphosphoramidothioic acid O,S-dimethyl ester)	1.0
31218-83-4	Propetamphos [3-[(Ethylamino) methoxyphosphinothioyl]oxy]-2-butenoic acid, 1-methylethyl ester]	1.0
33089-61-1	Amitraz	1.0
34014-18-1	Tebuthiuron [N-[5-(1,1-Dimethylethyl)-1,3,4-thiadiazol-2-yl]-N,N'-dimethylurea]	1.0
34077-87-7	Dichlorotrifluoroethane	1.0
35367-38-5	Diflubenzuron	1.0
35400-43-2	Sulprofos [O-Ethyl O-[4-(methylthio)phenyl]-phosphorodithioic acid S-propyl ester]	1.0

Table II. EPCRA Section 313 Chemical List for Reporting Year 2012

CAS Number	Chemical Name	De minimis % Limit
	Arranged by CAS Number	
35554-44-0	Imazalil [1-[2-(2,4-Dichlorophenyl)-2-(2-propenyloxy)ethyl]-1H-imidazole]	1.0
35691-65-7	1-Bromo-1-(bromomethyl)-1,3-propanedicarbonitrile	1.0
38727-55-8	Diethatyl ethyl	1.0
39156-41-7	2,4-Diaminoanisole sulfate	0.1
39300-45-3	Dinocap	1.0
39515-41-8	Fenpropathrin [2,2,3,3-Tetramethylcyclopropane carboxylic acid cyano(3-phenoxyphenyl)methyl ester]	1.0
40487-42-1	Pendimethalin [N-(1-Ethylpropyl)-3,4-dimethyl-2,6-dinitrobenzenamine]	*
41198-08-7	Profenofos [O-(4-Bromo-2-chlorophenyl)-O-ethyl-S-propyl phosphorothioate]	1.0
41766-75-0	3,3'-Dimethylbenzidine dihydrofluoride (o-Tolidinedihydrofluoride)	0.1
42874-03-3	Oxyfluorfen	1.0
43121-43-3	Triadimefon [1-(4-Chlorophenoxy)-3,3-dimethyl-1-(1H-1,2,4-triazol-1-yl)-2-butanone]	1.0
50471-44-8	Vinclozolin [3-(3,5-Dichlorophenyl)-5-ethenyl-5-methyl-2,4-oxazolidinedione]	1.0
51235-04-2	Hexazinone	1.0
51338-27-3	Diclofop methyl [2-[4-(2,4-Dichlorophenoxy)-phenoxy]propanoic acid, methyl ester]	1.0
51630-58-1	Fenvalerate [4-Chloro-alpha-(1-methylethyl)-benzeneacetic acid cyano(3-phenoxyphenyl)methyl ester]	1.0
52645-53-1	Permethrin [3-(2,2-Dichloroethenyl)-2,2-dimethylcyclopropane carboxylic acid, (3-phenoxyphenyl)methyl ester]	1.0
53404-19-6	Bromacil, lithium salt [2,4(1H,3H)-Pyrimidinedione, 5-bromo-6-methyl-3-(1-methylpropyl), lithium salt]	1.0
53404-37-8	2,4-D 2-ethyl-4-methylpentyl ester	0.1
53404-60-7	Dazomet, sodium salt [Tetrahydro-3,5-dimethyl-2H-1,3,5-thiadiazine-2-thione, ion(1-), sodium]	1.0
55290-64-7	Dimethipin [2,3-Dihydro-5,6-dimethyl-1,4-dithiin 1,1,4,4-tetraoxide]	1.0
55406-53-6	3-Iodo-2-propynyl butyl carbamate	1.0
57213-69-1	Triclopyr triethylammonium salt	1.0
59669-26-0	Thiodicarb	1.0
60168-88-9	Fenarimol [.alpha.-(2-Chlorophenyl)-.alpha.-(4-	1.0

CAS Number	Chemical Name	De minimis % Limit
	Arranged by CAS Number	
	chlorophenyl)-5-pyrimidinemethanol]	
60207-90-1	Propiconazole [1-[2-(2,4-Dichlorophenyl)-4-propyl-1,3-dioxolan-2-yl]-methyl-1H-1,2,4,-triazole]	1.0
62476-59-9	Acifluorfen, sodium salt [5-(2-Chloro-4-(trifluoromethyl)phenoxy)-2-nitrobenzoic acid, sodium salt]	1.0
63938-10-3	Chlorotetrafluoroethane	1.0
64902-72-3	Chlorsulfuron [2-Chloro-N-[[(4-methoxy-6-methyl-1,3,5-triazin-2-yl)amino] carbonyl] benzenesulfonamide]	1.0
64969-34-2	3,3'-Dichlorobenzidine sulfate	0.1
66441-23-4	Fenoxaprop ethyl [2-(4-((6-Chloro-2-benzoxazolylen)oxy)phenoxy)propanoic acid, ethyl ester]	1.0
67485-29-4	Hydramethylnon [Tetrahydro-5,5-dimethyl-2(1H)-pyrimidinone[3-[4-(trifluoromethyl)phenyl]-1-[2-[4-(trifluoromethyl)phenyl]ethenyl]-2-propenylidene]hydrazone]	1.0
68085-85-8	Cyhalothrin [3-(2-Chloro-3,3,3-trifluoro-1-propenyl)-2,2-dimethylcyclopropanecarboxylic acid cyano(3-phenoxyphenyl) methyl ester]	1.0
68359-37-5	Cyfluthrin [3-(2,2-Dichloroethenyl)-2,2-dimethylcyclopropanecarboxylic acid, cyano(4-fluoro-3-phenoxyphenyl) methyl ester]	1.0
69409-94-5	Fluvalinate [N-[2-Chloro-4-(trifluoromethyl)phenyl]DL-valine(+)-cyano(3-phenoxyphenyl)methyl ester]	1.0
69806-50-4	Fluazifop butyl [2-[4-[[5-(Trifluoromethyl)-2-pyridinyl]oxy]phenoxy]propanoic acid, butyl ester]	1.0
71751-41-2	Abamectin [Avermectin B1]	1.0
72178-02-0	Fomesafen [5-(2-Chloro-4-(trifluoromethyl)phenoxy)-N-methylsulfonyl)-2-nitrobenzamide]	1.0
72490-01-8	Fenoxycarb [[2-(4-Phenoxy phenoxy)ethyl]carbamic acid ethyl ester]	1.0
74051-80-2	Sethoxydim [2-[1-(Ethoxyimino)butyl]-5-[2-(ethylthio)propyl]-3-hydroxyl-2-cyclohexen-1-one]	1.0
76578-14-8	Quizalofop-ethyl [2-[4-[(6-Chloro-2-quinoxalinyl) oxy]phenoxy]propanoic acid ethyl ester]	1.0
77501-63-4	Lactofen [Benzoic acid, 5-[2-Chloro-4-(trifluoromethyl)phenoxy]-2-nitro-, 2-	1.0

Table II. EPCRA Section 313 Chemical List for Reporting Year 2012

CAS Number	Chemical Name	*De minimis* % Limit
	Arranged by CAS Number	
	ethoxy-1-methyl-2-oxoethyl ester]	
82657-04-3	Bifenthrin	1.0
88671-89-0	Myclobutanil	1.0
	[.alpha.-Butyl-.alpha.-(4-chlorophenyl)-1H-1,2,4-triazole-1-propanenitrile]	
90454-18-5	Dichloro-1,1,2-trifluoroethane	1.0
90982-32-4	Chlorimuron ethyl	1.0
	[Ethyl-2-[[[[(4-chloro-6-methoxyprimidin-2-yl)amino]carbonyl]amino]sulfonyl]benzoate]	
101200-48-0	Tribenuron methyl	1.0
	[2-[[[[(4-Methoxy-6-methyl-1,3,5-triazin-2-yl)methylamino]carbonyl]amino]sulfonyl]benzoic acid methyl ester]	
111512-56-2	1,1-Dichloro-1,2,3,3,3-pentafluoropropane (HCFC-225eb)	1.0
111984-09-9	3,3'-Dimethoxybenzidine hydrochloride (o-Dianisidine hydrochloride)	0.1
127564-92-5	Dichloropentafluoropropane	1.0
128903-21-9	2,2-Dichloro-1,1,1,3,3-pentafluoropropane (HCFC-225aa)	1.0
136013-79-1	1,3-Dichloro-1,1,2,3,3-pentafluoropropane (HCFC-225ea)	1.0

c.   Chemical Categories

Section 313 requires reporting on the EPCRA Section 313 chemical categories listed below, in addition to the specific EPCRA Section 313 chemicals listed above.

The metal compound categories listed below, unless otherwise specified, are defined as including any unique chemical substance that contains the named metal (e.g., antimony, nickel, etc.) as part of that chemical's structure.

EPCRA Section 313 chemical categories are subject to the 1% *de minimis* concentration unless the substance involved meets the definition of an OSHA carcinogen in which case the 0.1% *de minimis* concentration applies. The *de minimis* concentration for each category is provided in parentheses. The *de minimis* exemption is not available for PBT chemicals, therefore an asterisk appears where a *de minimis* limit would otherwise appear. However, for purposes of the supplier notification requirement only, such limits are provided in Appendix D.

N010   Antimony Compounds (1.0)
Includes any unique chemical substance that contains antimony as part of that chemical's infrastructure.

N020   Arsenic Compounds (inorganic compounds: 0.1; organic compounds: 1.0)
Includes any unique chemical substance that contains arsenic as part of that chemical's infrastructure.

N040   Barium Compounds (1.0)
Includes any unique chemical substance that contains barium as part of that chemical's infrastructure. This category does not include: Barium sulfate CAS Number 7727-43-7

N050   Beryllium Compounds (0.1)
Includes any unique chemical substance that contains beryllium as part of that chemical's infrastructure.

N078   Cadmium Compounds (0.1)
Includes any unique chemical substance that contains cadmium as part of that chemical's infrastructure.

N084   Chlorophenols (0.1)

Where x = 1 to 5

N090   Chromium Compounds (except for chromite ore mined in the Transvaal Region of South Africa and the unreacted ore component of the chromite ore processing residue (COPR). COPR is the solid waste remaining after aqueous extraction of oxidized chromite ore that has been combined with soda ash and kiln roasted at approximately 2,000 °F.) (chromium VI compounds: 0.1; chromium III compounds: 1.0)
Includes any unique chemical substance that contains chromium as part of that chemical's infrastructure.

N096   Cobalt Compounds (inorganic compounds: 0.1; organic compounds: 1.0)
Includes any unique chemical substance that contains cobalt as part of that chemical's infrastructure.

N100   Copper Compounds (1.0)
Includes any unique chemical substance that contains copper as part of that chemical's infrastructure. This category does not include copper phthalocyanine compounds that are substituted with only hydrogen, and/or chlorine, and/or bromine.

N106   Cyanide Compounds (1.0)
X^+CN^- *where X = H^+ or any other group where a formal dissociation can be made. For example KCN or Ca(CN)$_2$*

N120   Diisocyanates (1.0)
This category includes only those chemicals listed below.

CAS Number	Chemical Name
38661-72-2	1,3-Bis(methylisocyanate) - cyclohexane
10347-54-3	1,4-Bis(methylisocyanate)-cyclohexane
2556-36-7	1,4-Cyclohexane

Table II. EPCRA Section 313 Chemical List for Reporting Year 2012

	diisocyanate
134190-37-7	Diethyldiisocyanatobenzene
4128-73-8	4,4'-Diisocyanatodiphenyl ether
75790-87-3	2,4'-Diisocyanatodiphenyl sulfide
91-93-0	3,3'-Dimethoxybenzidine-4,4'-diisocyanate
91-97-4	3,3'-Dimethyl-4,4'-diphenylene diisocyanate
139-25-3	3,3'-Dimethyldiphenyl methane-4,4'-diisocyanate
822-06-0	Hexamethylene-1,6-diisocyanate
4098-71-9	Isophorone diisocyanate
75790-84-0	4-Methyldiphenylmethane-3,4-diisocyanate
5124-30-1	1,1-Methylenebis(4-isocyanatocyclohexane)
101-68-8	Methylenebis(phenylisocyanate) (MDI)
3173-72-6	1,5-Naphthalene diisocyanate
123-61-5	1,3-Phenylene diisocyanate
104-49-4	1,4-Phenylene diisocyanate
9016-87-9	Polymeric diphenylmethane diisocyanate
16938-22-0	2,2,4-Trimethylhexamethylene diisocyanate
15646-96-5	2,4,4-Trimethylhexamethylene diisocyanate

N150 **Dioxin and dioxin-like compounds (Manufacturing; and the processing or otherwise use of dioxin and dioxin-like compounds if the dioxin and dioxin-like compounds are present as contaminants in a chemical and if they were created during the manufacturing of that chemical.) (*)** This category includes only those chemicals listed below. [Note: When completing the Form R Schedule 1, enter the data for each member of the category in the order they are listed here (i.e., 1-17).]

Box #	CAS Number	Chemical Name
1	1746-01-6	2,3,7,8-Tetrachlorodibenzo-*p*-dioxin
2	40321-76-4	1,2,3,7,8-Pentachlorodibenzo-*p*-dioxin
3	39227-28-6	1,2,3,4,7,8-Hexachlorodibenzo-*p*-dioxin
4	57653-85-7	1,2,3,6,7,8-Hexachlorodibenzo-*p*-dioxin
5	19408-74-3	1,2,3,7,8,9-Hexachlorodibenzo-*p*-dioxin
6	35822-46-9	1,2,3,4,6,7,8-Heptachlorodibenzo-*p*-dioxin
7	3268-87-9	1,2,3,4,6,7,8,9-Octachlorodibenzo-*p*-dioxin
8	51207-31-9	2,3,7,8-Tetrachlorodibenzofuran
9	57117-41-6	1,2,3,7,8-Pentachlorodibenzofuran
10	57117-31-4	2,3,4,7,8-Pentachlorodibenzofuran
11	70648-26-9	1,2,3,4,7,8-Hexachlorod-benzofuran
12	57117-44-9	1,2,3,6,7,8-Hexachlorodibenzofuran
13	72918-21-9	1,2,3,7,8,9-Hexachlorodibenzofuran
14	60851-34-5	2,3,4,6,7,8-Hexachlorodibenzofuran
15	67562-39-4	1,2,3,4,6,7,8-Heptachlorodibenzofuran
16	55673-89-7	1,2,3,4,7,8,9-Heptachlorodibenzofuran
17	39001-02-0	1,2,3,4,6,7,8,9-Octachlorodibenzofuran

N171 **Ethylenebisdithiocarbamic acid, salts and esters EBDCs) (1.0)**
Includes any unique chemical substance that contains an EBDC or an EBDC salt as part of that chemical's infrastructure.

N230 **Certain Glycol Ethers (1.0)**

R - $(OCH_2CH_2)_n$ - OR'
where:
    n = 1, 2, or 3;
    R = Alkyl C7 or less; or
    R = phenyl or alkyl substituted phenyl;
    □' = H or alkyl C7 or less; or
    □□' consisting of carboxylic acid ester, sulfate, phosphate, nitrate, or sulfonate.

N420 **Lead Compounds (*)**
Includes any unique chemical substance that contains lead as part of that chemical's infrastructure.

Table II. EPCRA Section 313 Chemical List for Reporting Year 2012

N450 **Manganese Compounds (1.0)**
Includes any unique chemical substance that contains manganese as part of that chemical's infrastructure.

N458 **Mercury Compounds (*)**
Includes any unique chemical substance that contains mercury as part of that chemical's infrastructure.

N495 **Nickel Compounds (0.1)**
Includes any unique chemical substance that contains nickel as part of that chemical's infrastructure.

N503 **Nicotine and salts (1.0)**
Includes any unique chemical substance that contains nicotine or a nicotine salt as part of that chemical's infrastructure.

N511 **Nitrate compounds (water dissociable; reportable only when in aqueous solution) (1.0)**

N575 **Polybrominated Biphenyls (PBBs) (0.1)**

where x = 1 to 10

N583 **Polychlorinated alkanes (C_{10} to C_{13}) (1.0, except for those members of the category that have an average chain length of 12 carbons and contain an average chlorine content of 60% by weight which are subject to the 0.1% *de minimis*)**
Includes those chemicals defined by the following formula:

$$C_xH_{2x-y+2}Cl_y$$

Where x = 10 to 13;
  y = 3 to 12; and
  where the average chlorine content ranges from 40-70% with the limiting molecular formulas $C_{10}H_{19}Cl_3$ and $C_{13}H_{16}Cl_{12}$

N590 **Polycyclic aromatic compounds (PACs) (*)**
This category includes the chemicals listed below.

CAS Number	Chemical Name
56-55-3	Benz(a)anthracene
205-99-2	Benzo(b)fluoranthene
205-82-3	Benzo(j)fluoranthene
207-08-9	Benzo(k)fluoranthene
206-44-0	Benzo(j,k)fluorene
189-55-9	Benzo(r,s,t)pentaphene
218-01-9	Benzo(a)phenanthrene
50-32-8	Benzo(a)pyrene
226-36-8	Dibenz(a,h)acridine
224-42-0	Dibenz(a,j)acridine
53-70-3	Dibenzo(a,h)anthracene
194-59-2	7H-Dibenzo(c,g)carbazole

5385-75-1	Dibenzo(a,e)fluoranthene
192-65-4	Dibenzo(a,e)pyrene
189-64-0	Dibenzo(a,h)pyrene
191-30-0	Dibenzo(a,l)pyrene
57-97-6	7,12-Dimethylbenz(a)-anthracene
42397-64-8	1,6-Dinitropyrene
42397-65-9	1,8-Dinitropyrene
193-39-5	Indeno(1,2,3-cd)pyrene
56-49-5	3-Methylcholanthrene
3697-24-3	5-Methylchrysene
7496-02-8	6-Nitrochrysene
5522-43-0	1-Nitropyrene
57835-92-4	4-Nitropyrene

N725 **Selenium Compounds (1.0)**
Includes any unique chemical substance that contains selenium as part of that chemical's infrastructure.

N740 **Silver Compounds (1.0)**
Includes any unique chemical substance that contains silver as part of that chemical's infrastructure.

N746 **Strychnine and salts (1.0)**
Includes any unique chemical substance that contains strychnine or a strychnine salt as part of that chemical's infrastructure.

N760 **Thallium Compounds (1.0)**
Includes any unique chemical substance that contains

thallium as part of that chemical's infrastructure.

N770 **Vanadium compounds (1.0)**
Includes any unique chemical substance that contains vanadium as part of that chemical's infrastructure.

N874 **Warfarin and salts (1.0)**
Includes any unique chemical substance that contains warfarin or a warfarin salt as part of that chemical's infrastructure.

N982 **Zinc Compounds (1.0)**
Includes any unique chemical substance that contains zinc as part of that chemical's infrastructure.

Table III. State Abbreviations

Alabama	AL	Montana	MT
Alaska	AK	Nebraska	NE
American Samoa	AS	Nevada	NV
Arizona	AZ	New Hampshire	NH
Arkansas	AR	New Jersey	NJ
California	CA	New Mexico	NM
Colorado	CO	New York	NY
Connecticut	CT	North Carolina	NC
Delaware	DE	North Dakota	ND
District of Columbia	DC	Northern Marianas Islands	MP
Florida	FL	Ohio	OH
Georgia	GA	Oklahoma	OK
Guam	GU	Oregon	OR
Hawaii	HI	Pennsylvania	PA
Idaho	ID	Puerto Rico	PR
Illinois	IL	Rhode Island	RI
Indiana	IN	South Carolina	SC
Iowa	IA	South Dakota	SD
Kansas	KS	Tennessee	TN
Kentucky	KY	Texas	TX
Louisiana	LA	Utah	UT
Maine	ME	Vermont	VT
Marshall Islands	MH	Virginia	VA
Maryland	MD	Virgin Islands	VI
Massachusetts	MA	Washington	WA
Michigan	MI	West Virginia	WV
Minnesota	MN	Wisconsin	WI
Mississippi	MS	Wyoming	WY
Missouri	MO		

Table IV. Federal Information Processing Standards (FIPS) Country Codes

AA	Aruba	CE	Sri Lanka	FS	French Southern and
AC	Antigua and Barbuda	CF	Congo		Antarctic Lands
AE	United Arab		(Brazzaville)	GA	The Gambia
	Emirates	CG	Congo (Kinshasa)	GB	Gabon
AF	Afghanistan	CH	China	GG	Georgia
AG	Algeria	CI	Chile	GH	Ghana
AJ	Azerbaijan	CJ	Cayman Islands	GI	Gibraltar
AL	Albania	CK	Cocos (Keeling)	GJ	Grenada
AM	Armenia		Islands	GK	Guernsey
AN	Andorra	CM	Cameroon	GL	Greenland
AO	Angola	CN	Comoros	GM	Germany
AR	Argentina	CO	Colombia	GO	Glorioso Islands
AS	Australia	CR	Coral Sea Islands	GP	Guadeloupe
AT	Ashmore and Cartier	CS	Costa Rica	GR	Greece
	Islands	CT	Central African	GT	Guatemala
AU	Austria		Republic	GV	Guinea
AV	Anguilla	CU	Cuba	GY	Guyana
AY	Antarctica	CV	Cape Verde	GZ	Gaza Strip
BA	Bahrain	CW	Cook Islands	HA	Haiti
BB	Barbados	CY	Cyprus	HK	Hong Kong
BC	Botswana	DA	Denmark	HM	Heard Island and
BD	Bermuda	DJ	Djibouti		McDonald Islands
BE	Belgium	DO	Dominica	HO	Honduras
BF	The Bahamas	DR	Dominican	HR	Croatia
BG	Bangladesh		Republic	HU	Hungary
BH	Belize	EC	Ecuador	IC	Iceland
BK	Bosnia and	EG	Egypt	ID	Indonesia
	Herzegovina	EI	Ireland	IM	Isle of Man
BL	Bolivia	EK	Equatorial Guinea	IN	India
BM	Burma	EN	Estonia	IO	British Indian Ocean
BN	Benin	ER	Eritrea		Territory
BO	Belarus	ES	El Salvador	IP	Clipperton Island
BP	Solomon Islands	ET	Ethiopia	IR	Iran
BR	Brazil	EU	Europa Island	IS	Israel
BS	Bassas da India	EZ	Czech Republic	IT	Italy
BT	Bhutan	FG	French Guiana	IV	Cote D'Ivoire
BU	Bulgaria	FI	Finland	IZ	Iraq
BV	Bouvet Island	FJ	Fiji	JA	Japan
BX	Brunei	FK	Falkland Islands	JE	Jersey
BY	Burundi		(Islas Malvinas)	JM	Jamaica
CA	Canada	FO	Faroe Islands	JN	Jan Mayen
CB	Cambodia	FP	French Polynesia	JO	Jordan
CD	Chad	FR	France	JU	Juan de Nova Island
KE	Kenya	KS	South Korea	LE	Lebanon
KG	Kyrgyzstan	KT	Christmas Island	LG	Latvia
KN	North Korea	KU	Kuwait	LH	Lithuania
KQ	Kingman Reef	KZ	Kazakhstan	LI	Liberia
KR	Kiribati	LA	Laos	LO	Slovakia

Table IV. Federal Information Processing Standards (FIPS) Country Codes

LS	Liechtenstein	PF	Paracel Islands	TH	Thailand
LT	Lesotho	PG	Spratly Islands	TI	Tajikistan
LU	Luxembourg	PK	Pakistan	TK	Turks and Caicos Islands
LY	Libya	PL	Poland		
MA	Madagascar	PM	Panama	TL	Tokelau
MB	Martinique	PO	Portugal	TN	Tonga
MC	Macau	PP	Papua New Guinea	TO	Togo
MD	Moldova	PS	Palau	TP	Sao Tome and Principe
MF	Mayotte	PU	Guinea-Bissau		
MG	Mongolia	QA	Qatar	TS	Tunisia
MH	Montserrat	RE	Reunion	TT	East Timor
MI	Malawi	RO	Romania	TU	Turkey
MK	Macedonia	RP	Philippines	TV	Tuvalu
ML	Mali	RS	Russia	TW	Taiwan
MN	Monaco	RW	Rwanda	TX	Turkmenistan
MO	Morocco	SA	Saudi Arabia	TZ	Tanzania
MP	Mauritius	SB	St. Pierre and Miquelon	UG	Uganda
MR	Mauritania			UK	United Kingdom
MT	Malta	SC	St. Kitts and Nevis	UP	Ukraine
MU	Oman	SE	Seychelles	UV	Burkina Faso
MV	Maldives	SF	South Africa	UY	Uruguay
MX	Mexico	SG	Senegal	UZ	Uzbekistan
MY	Malaysia	SH	St. Helena	VC	St. Vincent and the Grenadines
MZ	Mozambique	SI	Slovenia		
NC	New Caledonia	SL	Sierra Leone	VE	Venezuela
NE	Niue	SM	San Marino	VI	British Virgin Islands
NF	Norfolk Island	SN	Singapore		
NG	Niger	SO	Somalia	VM	Vietnam
NH	Vanuatu	SP	Spain	VT	Vatican City
NI	Nigeria	ST	St. Lucia	WA	Namibia
NL	Netherlands	SU	Sudan	WE	West Bank
NO	Norway	SV	Svalbard	WF	Wallis and Futuna
NP	Nepal	SW	Sweden	WI	Western Sahara
NR	Nauru	SX	South Georgia and South Sandwich Islands	WS	Western Samoa
NS	Suriname			WZ	Swaziland
NT	Netherlands Antilles			YI	Yugoslavia
NU	Nicaragua	SY	Syria	YM	Yemen
NZ	New Zealand	SZ	Switzerland	ZA	Zambia
PA	Paraguay	TD	Trinidad and Tobago	ZI	Zimbabwe
PC	Pitcairn Islands				
PE	Peru	TE	Tromelin Island		

Table V. Bureau of Indian Affairs (BIA) Tribal Codes

Indian Country Name	BIA Tribe Code
Absentee-Shawnee Tribe of Indians of Oklahoma	820
Agua Caliente Band of Cahuilla Indians of the Agua Caliente Indian Reservation, California	584
Ak Chin Indian Community of the Maricopa (Ak Chin) Indian Reservation, Arizona	612
Alabama-Coushatta Tribes of Texas	830
Alabama-Quassarte Tribal Town, Oklahoma	901
Alturas Indian Rancheria, California	502
Apache Tribe of Oklahoma	809
Arapahoe Tribe of the Wind River Reservation, Wyoming	281
Aroostook Band of Micmac Indians of Maine	31
Assiniboine and Sioux Tribes of the Fort Peck Indian Reservation, Montana	206
Augustine Band of Cahuilla Indians, California (formerly the Augustine Band of Cahuilla Mission Indians of the Augustine Reservation)	567
Bad River Band of the Lake Superior Tribe of Chippewa Indians of the Bad River Reservation, Wisconsin	430
Bay Mills Indian Community, Michigan	470
Bear River Band of the Rohnerville Rancheria, California	560
Berry Creek Rancheria of Maidu Indians of California	504
Big Lagoon Rancheria, California	554
Big Pine Band of Owens Valley Paiute Shoshone Indians of the Big Pine Reservation, California	530
Big Sandy Rancheria of Mono Indians of California	506
Big Valley Band of Pomo Indians of the Big Valley Rancheria, California	507
Blackfeet Tribe of the Blackfeet Indian Reservation of Montana	201
Blue Lake Rancheria, California	558
Bridgeport Paiute Indian Colony of California	691
Buena Vista Rancheria of Me-Wuk Indians of California	508
Burns Paiute Tribe of the Burns Paiute Indian Colony of Oregon	144
Cabazon Band of Mission Indians, California	568
Cachil DeHe Band of Wintun Indians of the Colusa Indian Community of the Colusa Rancheria, California	512
Caddo Nation of Oklahoma	806
Cahto Indian Tribe of the Laytonville Rancheria, California	524

Indian Country Name	BIA Tribe Code
Cahuilla Band of Mission Indians of the Cahuilla Reservation, California	569
California Valley Miwok Tribe, California	628
Campo Band of Diegueno Mission Indians of the Campo Indian Reservation, California	570
Capitan Grande Band of Diegueno Mission Indians of California: Barona Group of Capitan Grande Band of Mission Indians of the Barona Reservation, California; Viejas (Baron Long) Group of Capitan Grande Band of Mission Indians of the Viejas Reservation, California	571
Catawba Indian Nation (aka Catawba Tribe of South Carolina)	32
Cayuga Nation of New York	13
Cedarville Rancheria, California	621
Chemehuevi Indian Tribe of the Chemehuevi Reservation, California	695
Cher-Ae Heights Indian Community of the Trinidad Rancheria, California	566
Cherokee Nation, Oklahoma	905
Cheyenne and Arapaho Tribes, Oklahoma (formerly the Cheyenne-Arapaho Tribes of Oklahoma)	801
Cheyenne River Sioux Tribe of the Cheyenne River Reservation, South Dakota	340
Chickasaw Nation, Oklahoma	906
Chicken Ranch Rancheria of Me-Wuk Indians of California	523
Chippewa-Cree Indians of the Rocky Boy's Reservation, Montana	205
Chitimacha Tribe of Louisiana	970
Choctaw Nation of Oklahoma	907
Citizen Potawatomi Nation, Oklahoma	821
Cloverdale Rancheria of Pomo Indians of California	510
Cocopah Tribe of Arizona	602
Coeur D'Alene Tribe of the Coeur D'Alene Reservation, Idaho	181
Cold Springs Rancheria of Mono Indians of California	511
Colorado River Indian Tribes of the Colorado River Indian Reservation, Arizona and California	603
Comanche Nation, Oklahoma	808
Confederated Salish & Kootenai Tribes of the Flathead Reservation, Montana	203
Confederated Tribes and Bands of the Yakama Nation, Washington	124

Table V. Bureau of Indian Affairs (BIA) Tribal Codes

Indian Country Name	BIA Tribe Code
Confederated Tribes of Siletz Indians of Oregon (previously listed as the Confederated Tribes of the Siletz Reservation)	142
Confederated Tribes of the Chehalis Reservation, Washington	105
Confederated Tribes of the Colville Reservation, Washington	101
Confederated Tribes of the Coos, Lower Umpqua and Siuslaw Indians of Oregon	152
Confederated Tribes of the Goshute Reservation, Nevada and Utah	681
Confederated Tribes of the Grand Ronde Community of Oregon	141
Confederated Tribes of the Umatilla Reservation, Oregon	143
Confederated Tribes of the Warm Springs Reservation of Oregon	145
Coquille Tribe of Oregon	155
Cortina Indian Rancheria of Wintun Indians of California	513
Coushatta Tribe of Louisiana	971
Cow Creek Band of Umpqua Indians of Oregon	153
Cowlitz Indian Tribe, Washington	132
Coyote Valley Band of Pomo Indians of California	638
Crow Creek Sioux Tribe of the Crow Creek Reservation, South Dakota	342
Crow Tribe of Montana	202
Death Valley Timbi-Sha Shoshone Band of California	693
Delaware Nation, Oklahoma	807
Delaware Tribe of Indians, Oklahoma	816
Dry Creek Rancheria of Pomo Indians of California	515
Duckwater Shoshone Tribe of the Duckwater Reservation, Nevada	642
Eastern Band of Cherokee Indians of North Carolina	1
Eastern Shawnee Tribe of Oklahoma	921
Elem Indian Colony of Pomo Indians of the Sulphur Bank Rancheria, California	632
Elk Valley Rancheria, California	559
Ely Shoshone Tribe of Nevada	644
Enterprise Rancheria of Maidu Indians of California	517
Ewiiaapaayp Band of Kumeyaay Indians, California	573
Federated Indians of Graton Rancheria, California	622
Flandreau Santee Sioux Tribe of South Dakota	341
Forest County Potawatomi Community, Wisconsin	434
Fort Belknap Indian Community of the Fort Belknap Reservation of Montana	204

Indian Country Name	BIA Tribe Code
Fort Bidwell Indian Community of the Fort Bidwell Reservation of California	518
Fort Independence Indian Community of Paiute Indians of the Fort Independence Reservation, California	525
Fort McDermitt Paiute and Shoshone Tribes of the Fort McDermitt Indian Reservation, Nevada and Oregon	646
Fort McDowell Yavapai Nation, Arizona	613
Fort Mojave Indian Tribe of Arizona, California & Nevada	604
Fort Sill Apache Tribe of Oklahoma	803
Gila River Indian Community of the Gila River Indian Reservation, Arizona	614
Grand Traverse Band of Ottawa and Chippewa Indians, Michigan	468
Greenville Rancheria of Maidu Indians of California	545
Grindstone Indian Rancheria of Wintun-Wailaki Indians of California	519
Habematolel Pomo of Upper Lake, California	636
Hannahville Indian Community, Michigan	471
Havasupai Tribe of the Havasupai Reservation, Arizona	605
Ho-Chunk Nation of Wisconsin	439
Hoh Indian Tribe of the Hoh Indian Reservation, Washington	106
Hoopa Valley Tribe, California	561
Hopi Tribe of Arizona	608
Hopland Band of Pomo Indians of the Hopland Rancheria, California	521
Houlton Band of Maliseet Indians of Maine	19
Hualapai Indian Tribe of the Hualapai Indian Reservation, Arizona	606
Iipay Nation of Santa Ysabel, California (formerly the Santa Ysabel Band of Diegueno Mission Indians of the Santa Ysabel Reservation)	592
Inaja Band of Diegueno Mission Indians of the Inaja and Cosmit Reservation, California	574
Ione Band of Miwok Indians of California	529
Iowa Tribe of Kansas and Nebraska	860
Iowa Tribe of Oklahoma	822
Jackson Rancheria of Me-Wuk Indians of California	522
Jamestown S'Klallam Tribe of Washington	129
Jamul Indian Village of California	575
Jena Band of Choctaw Indians, Louisiana	34
Jicarilla Apache Nation, New Mexico	701

Table V. Bureau of Indian Affairs (BIA) Tribal Codes

Indian Country Name	BIA Tribe Code
Kaibab Band of Paiute Indians of the Kaibab Indian Reservation, Arizona	617
Kalispel Indian Community of the Kalispel Reservation, Washington	103
Karuk Tribe (formerly the Karuk Tribe of California)	555
Kashia Band of Pomo Indians of the Stewarts Point Rancheria, California	547
Kaw Nation, Oklahoma	810
Kewa Pueblo, New Mexico (formerly the Pueblo of Santo Domingo)	717
Keweenaw Bay Indian Community, Michigan	475
Kialegee Tribal Town, Oklahoma	902
Kickapoo Traditional Tribe of Texas	826
Kickapoo Tribe of Indians of the Kickapoo Reservation in Kansas	861
Kickapoo Tribe of Oklahoma	823
Kiowa Indian Tribe of Oklahoma	802
Klamath Tribes, Oregon	140
Kootenai Tribe of Idaho	183
La Jolla Band of Luiseno Indians, California (formerly the La Jolla Band of Luiseno Mission Indians of the La Jolla Reservation)	576
La Posta Band of Diegueno Mission Indians of the La Posta Indian Reservation, California	577
Lac Courte Oreilles Band of Lake Superior Chippewa Indians of Wisconsin	431
Lac du Flambeau Band of Lake Superior Chippewa Indians of the Lac du Flambeau Reservation of Wisconsin	432
Lac Vieux Desert Band of Lake Superior Chippewa Indians, Michigan	479
Las Vegas Tribe of Paiute Indians of the Las Vegas Indian Colony, Nevada	648
Little River Band of Ottawa Indians, Michigan	482
Little Traverse Bay Bands of Odawa Indians, Michigan	483
Los Coyotes Band of Cahuilla and Cupeno Indians, California (formerly the Los Coyotes Band of Cahuilla & Cupeno Indians of the Los Coyotes Reservation)	578
Lovelock Paiute Tribe of the Lovelock Indian Colony, Nevada	649
Lower Brule Sioux Tribe of the Lower Brule Reservation, South Dakota	343
Lower Elwha Tribal Community of the Lower Elwha Reservation, Washington	125
Lower Lake Rancheria, California	625

Indian Country Name	BIA Tribe Code
Lower Sioux Indian Community in the State of Minnesota	402
Lummi Tribe of the Lummi Reservation, Washington	107
Lytton Rancheria of California	509
Makah Indian Tribe of the Makah Indian Reservation, Washington	108
Manchester Band of Pomo Indians of the Manchester-Point Arena Rancheria, California	527
Manzanita Band of Diegueno Mission Indians of the Manzanita Reservation, California	579
Mashantucket Pequot Tribe of Connecticut	20
Mashpee Wampanoag Tribe, Massachusetts	35
Match-e-be-nash-she-wish Band of Pottawatomi Indians of Michigan	484
Mechoopda Indian Tribe of Chico Rancheria, California	531
Menominee Indian Tribe of Wisconsin	440
Mesa Grande Band of Diegueno Mission Indians of the Mesa Grande Reservation, California	580
Mescalero Apache Tribe of the Mescalero Reservation, New Mexico	702
Miami Tribe of Oklahoma	925
Miccosukee Tribe of Indians of Florida	26
Middletown Rancheria of Pomo Indians of California	528
Minnesota Chippewa Tribe, Minnesota (Six component reservations: Bois Forte Band (Nett Lake); Fond du Lac Band; Grand Portage Band; Leech Lake Band; Mille Lacs Band; White Earth Band)	400
Mississippi Band of Choctaw Indians, Mississippi	980
Moapa Band of Paiute Indians of the Moapa River Indian Reservation, Nevada	650
Modoc Tribe of Oklahoma	927
Mohegan Indian Tribe of Connecticut	33
Mooretown Rancheria of Maidu Indians of California	626
Morongo Band of Mission Indians, California (formerly the Morongo Band of Cahuilla Mission Indians of the Morongo Reservation)	582
Muckleshoot Indian Tribe of the Muckleshoot Reservation, Washington	109
Muscogee (Creek) Nation, Oklahoma	908
Narragansett Indian Tribe of Rhode Island	27
Navajo Nation, Arizona, New Mexico & Utah	780
Nez Perce Tribe, Idaho (previously listed as Nez Perce Tribe of Idaho)	182
Nisqually Indian Tribe of the Nisqually Reservation, Washington	110

Table V. Bureau of Indian Affairs (BIA) Tribal Codes

Indian Country Name	BIA Tribe Code
Nooksack Indian Tribe of Washington	111
Northern Cheyenne Tribe of the Northern Cheyenne Indian Reservation, Montana	207
Northfork Rancheria of Mono Indians of California	532
Northwestern Band of Shoshoni Nation of Utah (Washakie)	195
Nottawaseppi Huron Band of the Potawatomi, Michigan (formerly the Huron Potawatomi, Inc.)	481
Oglala Sioux Tribe of the Pine Ridge Reservation, South Dakota	344
Ohkay Owingeh, New Mexico (formerly the Pueblo of San Juan)	714
Omaha Tribe of Nebraska	380
Oneida Nation of New York	11
Oneida Tribe of Indians of Wisconsin	433
Onondaga Nation of New York	6
Osage Nation, Oklahoma (formerly the Osage Tribe)	930
Otoe-Missouria Tribe of Indians, Oklahoma	811
Ottawa Tribe of Oklahoma	922
Paiute Indian Tribe of Utah (Cedar Band of Paiutes, Kanosh Band of Paiutes, Koosharem Band of Paiutes, Indian Peaks Band of Paiutes, and Shivwits Band of Paiutes) (formerly Paiute Indian Tribe of Utah (Cedar City Band of Paiutes, Kanosh Band of Paiutes, Koosharem Band of Paiutes, Indian Peaks Band of Paiutes, and Shivwits Band of Paiutes))	692
Paiute-Shoshone Indians of the Bishop Community of the Bishop Colony, California	549
Paiute-Shoshone Indians of the Lone Pine Community of the Lone Pine Reservation, California	624
Paiute-Shoshone Tribe of the Fallon Reservation and Colony, Nevada	645
Pala Band of Luiseno Mission Indians of the Pala Reservation, California	583
Pascua Yaqui Tribe of Arizona	665
Paskenta Band of Nomlaki Indians of California	533
Passamaquoddy Tribe of Maine	14
Pauma Band of Luiseno Mission Indians of the Pauma & Yuima Reservation, California	585
Pawnee Nation of Oklahoma	812
Pechanga Band of Luiseno Mission Indians of the Pechanga Reservation, California	586
Penobscot Tribe of Maine	18
Peoria Tribe of Indians of Oklahoma	926
Picayune Rancheria of Chukchansi Indians of California	534
Pinoleville Pomo Nation, California (formerly the Pinoleville Rancheria of Pomo Indians of California)	535

Indian Country Name	BIA Tribe Code
Pit River Tribe, California (includes XL Ranch, Big Bend, Likely, Lookout, Montgomery Creek and Roaring Creek Rancherias)	536
Pokagon Band of Potawatomi Indians, Michigan and Indiana	480
Ponca Tribe of Indians of Oklahoma	813
Ponca Tribe of Nebraska	381
Port Gamble Indian Community of the Port Gamble Reservation, Washington	113
Potter Valley Tribe, California	537
Prairie Band of Potawatomi Nation, Kansas	862
Prairie Island Indian Community in the State of Minnesota	403
Pueblo of Acoma, New Mexico	703
Pueblo of Cochiti, New Mexico	704
Pueblo of Isleta, New Mexico	705
Pueblo of Jemez, New Mexico	706
Pueblo of Laguna, New Mexico	707
Pueblo of Nambe, New Mexico	708
Pueblo of Picuris, New Mexico	709
Pueblo of Pojoaque, New Mexico	710
Pueblo of San Felipe, New Mexico	712
Pueblo of San Ildefonso, New Mexico	713
Pueblo of Sandia, New Mexico	711
Pueblo of Santa Ana, New Mexico	715
Pueblo of Santa Clara, New Mexico	716
Pueblo of Taos, New Mexico	718
Pueblo of Tesuque, New Mexico	719
Pueblo of Zia, New Mexico	720
Puyallup Tribe of the Puyallup Reservation, Washington	115
Pyramid Lake Paiute Tribe of the Pyramid Lake Reservation, Nevada	651
Quapaw Tribe of Indians, Oklahoma	920
Quartz Valley Indian Community of the Quartz Valley Reservation of California	563
Quechan Tribe of the Fort Yuma Indian Reservation, California & Arizona	696
Quileute Tribe of the Quileute Reservation, Washington	116
Quinault Tribe of the Quinault Reservation, Washington	117
Ramona Band of Cahuilla, California (formerly the Ramona Band or Village of Cahuilla Mission Indians of California)	597

Table V. Bureau of Indian Affairs (BIA) Tribal Codes

Indian Country Name	BIA Tribe Code
Red Cliff Band of Lake Superior Chippewa Indians of Wisconsin	435
Red Lake Band of Chippewa Indians, Minnesota	409
Redding Rancheria, California	538
Redwood Valley Rancheria of Pomo Indians of California	539
Reno-Sparks Indian Colony, Nevada	653
Resighini Rancheria, California	556
Rincon Band of Luiseno Mission Indians of the Rincon Reservation, California	587
Robinson Rancheria of Pomo Indians of California	516
Rosebud Sioux Tribe of the Rosebud Indian Reservation, South Dakota	345
Round Valley Indian Tribes of the Round Valley Reservation, California	540
Sac & Fox Nation of Missouri in Kansas and Nebraska	863
Sac & Fox Nation, Oklahoma	824
Sac & Fox Tribe of the Mississippi in Iowa	490
Saginaw Chippewa Indian Tribe of Michigan	472
Saint Regis Mohawk Tribe, New York (formerly the St. Regis Band of Mohawk Indians of New York)	7
Salt River Pima-Maricopa Indian Community of the Salt River Reservation, Arizona	615
Samish Indian Tribe, Washington	133
San Carlos Apache Tribe of the San Carlos Reservation, Arizona	616
San Juan Southern Paiute Tribe of Arizona	689
San Manuel Band of Mission Indians, California (previously listed as the San Manual Band of Serrano Mission Indians of the San Manual Reservation)	588
San Pasqual Band of Dieguano Mission Indians of California	589
Santa Rosa Band of Cahuilla Indians, California (formerly the Santa Rosa Band of Cahuilla Mission Indians of the Santa Rosa Reservation)	590
Santa Rosa Indian Community of the Santa Rosa Rancheria, California	542
Santa Ynez Band of Chumash Mission Indians of the Santa Ynez Reservation, California	591
Santee Sioux Nation, Nebraska	382
Sauk-Suiattle Indian Tribe of Washington	119
Sault Ste. Marie Tribe of Chippewa Indians of Michigan	469
Scotts Valley Band of Pomo Indians of California	503
Seminole Nation of Oklahoma	909

Indian Country Name	BIA Tribe Code
Seminole Tribe of Florida (Dania, Big Cypress, Brighton, Hollywood & Tampa Reservations)	21
Seneca Nation of New York	12
Seneca-Cayuga Tribe of Oklahoma	923
Shakopee Mdewakanton Sioux Community of Minnesota	411
Shawnee Tribe, Oklahoma	911
Sherwood Valley Rancheria of Pomo Indians of California	629
Shingle Springs Band of Miwok Indians, Shingle Springs Rancheria (Verona Tract), California	546
Shoalwater Bay Tribe of the Shoalwater Bay Indian Reservation, Washington	118
Shoshone Tribe of the Wind River Reservation, Wyoming	282
Shoshone-Bannock Tribes of the Fort Hall Reservation of Idaho	180
Shoshone-Paiute Tribes of the Duck Valley Reservation, Nevada	641
Sisseton-Wahpeton Oyate of the Lake Traverse Reservation, South Dakota	347
Skokomish Indian Tribe of the Skokomish Reservation, Washington	120
Skull Valley Band of Goshute Indians of Utah	682
Smith River Rancheria, California	564
Snoqualmie Tribe, Washington	126
Soboba Band of Luiseno Indians, California	593
Sokaogon Chippewa Community, Wisconsin	437
Southern Ute Indian Tribe of the Southern Ute Reservation, Colorado	750
Spirit Lake Tribe, North Dakota	303
Spokane Tribe of the Spokane Reservation, Washington	102
Squaxin Island Tribe of the Squaxin Island Reservation, Washington	121
St. Croix Chippewa Indians of Wisconsin	436
Standing Rock Sioux Tribe of North & South Dakota	302
Stillaguamish Tribe of Washington	139
Stockbridge Munsee Community, Wisconsin	438
Summit Lake Paiute Tribe of Nevada	655
Suquamish Indian Tribe of the Port Madison Reservation, Washington	114
Susanville Indian Rancheria, California	550
Swinomish Indians of the Swinomish Reservation, Washington	122
Sycuan Band of the Kumeyaay Nation	594
Table Mountain Rancheria of California	551

Table V. Bureau of Indian Affairs (BIA) Tribal Codes

Indian Country Name	BIA Tribe Code
Te-Moak Tribe of Western Shoshone Indians of Nevada (Four constituent bands: Battle Mountain Band; Elko Band; South Fork Band and Wells Band)	640
Thlopthlocco Tribal Town, Oklahoma	903
Three Affiliated Tribes of the Fort Berthold Reservation, North Dakota	301
Tohono O'odham Nation of Arizona	610
Tonawanda Band of Seneca Indians of New York	8
Tonkawa Tribe of Indians of Oklahoma	814
Tonto Apache Tribe of Arizona	674
Torres Martinez Desert Cahuilla Indians, California (formerly the Torres-Martinez Band of Cahuilla Mission Indians of California)	595
Tulalip Tribes of the Tulalip Reservation, Washington	123
Tule River Indian Tribe of the Tule River Reservation, California	553
Tunica-Biloxi Indian Tribe of Louisiana	336
Tuolumne Band of Me-Wuk Indians of the Tuolumne Rancheria of California	634
Turtle Mountain Band of Chippewa Indians of North Dakota	304
Tuscarora Nation of New York	9
Twenty-Nine Palms Band of Mission Indians of California	598
United Auburn Indian Community of the Auburn Rancheria of California	637
United Keetoowah Band of Cherokee Indians in Oklahoma	904
Upper Sioux Community, Minnesota	401
Upper Skagit Indian Tribe of Washington	131
Ute Indian Tribe of the Uintah & Ouray Reservation, Utah	687
Ute Mountain Tribe of the Ute Mountain Reservation, Colorado, New Mexico & Utah	751

Indian Country Name	BIA Tribe Code
Utu Utu Gwaitu Paiute Tribe of the Benton Paiute Reservation, California	520
Walker River Paiute Tribe of the Walker River Reservation, Nevada	656
Wampanoag Tribe of Gay Head (Aquinnah) of Massachusetts	30
Washoe Tribe of Nevada & California (Carson Colony, Dresslerville Colony, Woodfords Community, Stewart Community, & Washoe Ranches)	672
White Mountain Apache Tribe of the Fort Apache Reservation, Arizona	607
Wichita and Affiliated Tribes (Wichita, Keechi, Waco & Tawakonie), Oklahoma	804
Winnebago Tribe of Nebraska	383
Winnemucca Indian Colony of Nevada	659
Wiyot Tribe, California (formerly the Table Bluff Reservation—Wiyot Tribe)	565
Wyandotte Nation, Oklahoma	924
Yankton Sioux Tribe of South Dakota	346
Yavapai-Apache Nation of the Camp Verde Indian Reservation, Arizona	601
Yavapai-Prescott Tribe of the Yavapai Reservation, Arizona	618
Yerington Paiute Tribe of the Yerington Colony & Campbell Ranch, Nevada	660
Yocha Dehe Wintun Nation, California (formerly the Rumsey Indian Rancheria of Wintun Indians of California)	541
Yomba Shoshone Tribe of the Yomba Reservation, Nevada	661
Ysleta Del Sur Pueblo of Texas	725
Yurok Tribe of the Yurok Reservation, California	562
Zuni Tribe of the Zuni Reservation, New Mexico	721

Appendix A.   TRI Federal Facility Reporting Information

Special Instructions for TRI Federal Facility Reporting

A.1 Why Do Federal Facilities Need to Report?

Executive Order 13423, "Strengthening Federal Environmental Energy, and Transportation Management," requires federal agencies to comply with the Emergency Planning and Community Right-To-Know Act of 1986 (EPCRA) and the Pollution Prevention Act of 1990 (PPA). Federal facilities have been subject to EPCRA section 313 and PPA since reporting year 1994. TRI submissions are due to EPA on July 1 of the year following each reporting (calendar) year. Reporting by the federal facility does not alter the reporting obligation of on-site contractors. Contracts entered into after the date of this order for contractor operation of government-owned facilities or vehicles require the contractor to comply with the provisions of this order with respect to such facilities or vehicles to the same extent as the agency would be required to comply if the agency operated facilities or vehicles.

For more information on Executive Order 13423 please refer to the implementing instructions, which can be found on the TRI home page at http://www.epa.gov/tri

A.2 Identifying Federal Facility Reports

Federal facility reports are identified as federal by several indicators on the form. The facility name and parent company name are critical indicators and must be reported as described below. Another critical indicator is the federal facility report box, Part I, 4.2c. Federal facilities only should check this box to indicate that the report is from a federal agency for a federal facility; federal facilities should not check the GOCO box, (Part I, Section 4.2d of the Form R). Contractors located at federal facilities (GOCOs) should check the GOCO box (Part I, Section 4.2d of the Form R); they should not check the box 4.2c. Facilities should also complete the partial or complete facility blocks (Form R page 2, block 4.2a and 4.2b) as appropriate. If you are a federal facility reporting for the first time, you should write "new" in the TRI Facility ID (TRIFID) box, even if a contractor has reported for your facility in the past. The contractor will retain the original TRIFID. You will be assigned a new TRIFID the first time you report.

A.3 The "Double Counting" Problem

As structured, the law and the executive order require both regulated industries and the federal government to report TRI data, sometimes for the same site. In order to prevent duplicate data in the TRI database, which could result in "double counting" data for some chemicals and locations, EPA must be able to identify and distinguish the GOCO reports submitted by the federal contractor from the federal facility reports which contain data for the same site. To accomplish this, federal facility reports should be accompanied by either 1) exact copies (paper or electronic) of all contractor TRI reports, including when the totals reported by the federal facility are greater than that reported by the contractor(s), or 2) a cover letter which includes a list of the facility contractors which submit TRI reports to EPA, identifying each contractor by name, TRI technical contact, and TRI facility name and address. Additionally, federal facilities should check Form R, Part I, Section 4.2c, while contractors at federal facilities should check Form R, Part I, Section 4.2d.

A.4 How to Report Your Facility Name

Facility name is a critical data element. It is used by EPA to create the TRI facility ID number (TRIFID), which is a unique number designed to identify a facility site. The facility name and TRIFID number are used by all TRI data users to link data from a single site across multiple reporting years. A federal facility is assigned a new TRIFID number when the federal report is entered into the Toxics Release Inventory system for the first time. This TRIFID number, generated when the first report is entered into the Toxics Release Inventory System, will be included in future reporting packages sent to the federal facility, and should be used by the federal facility in all future reports.

Federal facilities should report their facility name on page 1 of the Form Rs (Section 4.1), as shown in the following example:

> U.S. DOE Savannah River Site

It is very important that the agency name appear first, followed by the specific plant or site name.

Federal contractors at GOCO facilities should report their names as shown in the following example:

> U.S. DOE Savannah River Site - Westinghouse Operations.

A.5 How to Report Your North American Industry Classification System (NAICS) Code

Federal facilities should report the NAICS code which most closely represents the activities taking place at the site. Additional guidance on determining your NAICS code is provided in the Forms and Instructions booklet. The table on the next page contains Public Administration NAICS codes covering executive, legislative, judicial, administrative and regulatory activities of the Federal government. Government-owned and operated business establishments are classified in major NAICS groups according to the activity in which they are engaged. For example, a Veterans Hospital would be classified in Group 806 - Hospitals.

A.6 How to Report Your "Parent Company" Name

Federal facilities should report their parent company name on page 2 of the Form Rs (Section 5.1) by reporting their complete Department or Agency name, as shown in the following example:

U.S. Department of Energy

Block 5.2, Parent Company's Dun & Bradstreet Number, should be marked NA.

Federal contractors at GOCO facilities should not report a federal department or agency name as their parent company. A federal name in the parent company name field will classify the report as federal, and the GOCO may be identified as a non-reporter.

A.7 How to Revise Your Data After It Has Been Submitted

Any TRI Form R submitter may voluntarily revise their submission if they find errors after their reports have been sent to EPA. If the revision is to a hardcopy report, the facility reporter should photocopy the original form and use a blue or black pen to mark out the incorrect value and write in the corrected value. The revised report should be submitted to EPA, with an "X" in the revision block on page 1 of the Form R. If the revision is to a diskette, a new diskette should be submitted, containing the data only for the revised submission, not all the chemicals originally reported. If a federal facility receives a copy of a revision from a contractor located at the federal facility, the facility should revise the federal

report, and submit the revised report to EPA and the appropriate state along with an exact copy of the contractor's revision. The cover letter from the federal facility should indicate that its submission is a revision.

A.8 Who Should Sign Federal Form R Reports?

Federal Form R reports should be signed by the senior federal employee on-site. If no federal employee is on-site, federal Form R reports must be signed by the senior federal employee with management responsibility for the site. Federal Form R reports should be signed by a federal employee. Contractor employee signatures are not considered valid on federal reports.

A.9 More Help is Available!

Federal facilities may call the EPA/TRI Information Center to ask specific questions concerning how to submit their Form R report. For contact information, see the TRI Home Page at http://www.epa.gov/tri/

A.10 North American Industry Classification System Codes 921-928

Sector 92 - Public Administration

921   Executive, Legislative, and Other General Government Support

92111	Executive Offices
92112	Legislative Bodies
92113	Public Finance Activities
92114	Executive and Legislative Offices Combined
92115	American Indian and Alaska Native Tribal Governments
92119	General Government, Not Elsewhere Classified

922   Justice, Public Order, and Safety Activities

92211	Courts
92212	Police Protection
92213	Legal Counsel and Prosecution
92214	Correctional Institutions
92215	Parole Offices and Probation Offices
92216	Fire Protection
92219	Other Justice, Public Order and Safety Activities

923 Administration of Human Resource Programs

92311   Administration of Educational Programs
92312   Administration of Public Health Programs
92313   Administration of Human Resource Programs (Except Education, Public Health, and Veterans' Affairs Programs)
92314   Administration of Veterans Affairs

924 Administration of Environmental Quality Programs

92411   Administration of Air and Water Resource and Solid Waste Management Programs
92412   Administration of Conservation Programs

925 Administration of Housing Programs, Urban Planning, and Community Development

92511   Administration of Housing Programs
92512   Administration of Urban Planning and Community and Rural Development

926 Administration of Economic Programs

92611   Administration of General Economic Programs
92612   Regulation and Administration of Transportation Programs
92613   Regulation and Administration of Communications, Electric, Gas, and Other Utilities
92614   Regulation of Agricultural Marketing and Commodities
92615   Regulation, Licensing, and Inspection of Miscellaneous Commercial Sectors

927 Space Research and Technology

92711   Space Research and Technology

928 National Security and International Affairs

92811   National Security
92812   International Affairs

Appendix B. Reporting Codes for EPA Form R and Instructions for Reporting Metals

B.1 Form R Part II

Revision Codes:

RR1   New Monitoring Data
RR2   New Emission Factor(s)
RR3   New Chemical Concentration Data
RR4   Recalculation(s)
RR5   Other Reason(s)

Withdrawal Codes:

WT1   Did not meet the reporting threshold for manufacturing, processing, or otherwise use
WT2   Did not meet the reporting threshold for number of employees
WT3   Not in a covered NAICS Code
WO1   Other reason(s)

Section 1.1.   CAS Number

EPCRA Section 313 Chemical Category Codes

N010   Antimony compounds
N020   Arsenic compounds
N040   Barium compounds
N050   Beryllium compounds
N078   Cadmium compounds
N084   Chlorophenols
N090   Chromium compounds
N096   Cobalt compounds
N100   Copper compounds
N106   Cyanide compounds
N120   Diisocyanates
N150   Dioxin and dioxin-like compounds
        N171Ethylenebisdithiocarbamic
        acid, salts and esters (EBDCs)
N230   Certain glycol ethers
N420   Lead compounds
N450   Manganese compounds
N458   Mercury compounds
N495   Nickel compounds
N503   Nicotine and salts
N511   Nitrate compounds
N575   Polybrominated biphenyls (PBBs)
N583   Polychlorinated alkanes
N590   Polycyclic aromatic compounds
N725   Selenium compounds
N740   Silver compounds
N746   Strychnine and salts
N760   Thallium compounds
N770   Vanadium compounds

N874   Warfarin and salts
N982   Zinc compounds

Section 4. Maximum Amount of the Toxic Chemical On-Site at Any Time During the Calendar Year

Range(pounds)

Range Code	From	To
01	0	99
02	100	999
03	1,000	9,999
04	10,000	99,999
05	100,000	999,999
06	1,000,000	9,999,999
07	10,000,000	49,999,999
08	50,000,000	99,999,999
09	100,000,000	499,999,999
10	500,000,000	999,999,999
11	1 billion	more than 1 billion

Section 5. Quantity of the Non-PBT Chemical Entering Each Environmental Medium On-site and Section 6. Transfers of the Toxic Chemical in Wastes to Off-Site Locations

Total Release or Transfer

Code	Range (pounds)
A	1-10
B	11-499
C	500-999

Basis of Estimate

M1-   Estimate is based on continuous monitoring data or measurements for the EPCRA section 313 chemical.

M2-   Estimate is based on periodic or random monitoring data or measurements for the EPCRA section 313 chemical.

C-   Estimate is based on mass balance calculations, such as calculation of the amount of the EPCRA section 313 chemical in streams entering and leaving process equipment.

E1- Estimate is based on published emission factors, such as those relating release quantity to through-put or equipment type (e.g., air emission factors).

E2- Estimate is based on site specific emission factors, such as those relating release quantity to through-put or equipment type (e.g., air emission factors).

O- Estimate is based on other approaches such as engineering calculations (e.g., estimating volatilization using published mathematical formulas) or best engineering judgment. This would include applying an estimated removal efficiency to a waste stream, even if the composition of the stream before treatment was fully identified through monitoring data.

Section 6. Transfers of the Toxic Chemical in Wastes to Off-Site Locations

Type of Waste Disposal/Treatment/Energy Recovery/Recycling

M10	Storage Only
M20	Solvents/Organics Recovery
M24	Metals Recovery
M26	Other Reuse or Recovery
M28	Acid Regeneration
M40	Solidification/Stabilization
M41	Solidification/Stabilization-Metals and Metal Category Compounds only
M50	Incineration/Thermal Treatment
M54	Incineration/Insignificant Fuel Value
M56	Energy Recovery
M61	Wastewater Treatment (Excluding POTW)
M62	Wastewater Treatment (Excluding POTW) - Metals and Metal Category Compounds only
M64	Other Landfills
M65	RCRA Subtitle C Landfills
M66	Subtitle C Surface Impoundment
M67	Other Surface Impoundments
M69	Other Waste Treatment
M73	Land Treatment
M79	Other Land Disposal
M81	Underground Injection to Class I Wells
M82	Underground Injection to Class II-V Wells
M90	Other Off-Site Management
M92	Transfer to Waste Broker - Energy Recovery
M93	Transfer to Waste Broker - Recycling
M94	Transfer to Waste Broker - Disposal
M95	Transfer to Waste Broker - Waste Treatment
M99	Unknown

Section 7A. On-Site Waste Treatment Methods and Efficiency

General Waste Stream

A	Gaseous (gases, vapors, airborne particulates)
W	Wastewater (aqueous waste)
L	Liquid waste streams (non-aqueous waste)
S	Solid waste streams (including sludges and slurries)

Waste Treatment Methods

Air Emissions Treatment

A01	Flare
A02	Condenser
A03	Scrubber
A04	Absorber
A05	Electrostatic Precipitator
A06	Mechanical Separation
A07	Other Air Emission Treatment

Chemical Treatment

H040	Incineration--thermal destruction other than use as a fuel
H071	Chemical reduction with or without precipitation
H073	Cyanide destruction with or without precipitation
H075	Chemical oxidation
H076	Wet air oxidation
H077	Other chemical precipitation with or without pre-treatment

Biological Treatment

H081	Biological treatment with or without precipitation

Physical Treatment

H082	Adsorption
H083	Air or steam stripping
H101	Sludge treatment and/or dewatering
H103	Absorption
H111	Stabilization or chemical fixation prior to disposal
H112	Macro-encapsulation prior to disposal
H121	Neutralization
H122	Evaporation
H123	Settling or clarification
H124	Phase separation
H129	Other treatment

Section 7B. On-Site Energy Recovery Processes

U01	Industrial Kiln
U02	Industrial Furnace

U03   Industrial Boiler

Section 7C. On-Site Recycling Processes

H10   Metal recovery (by retorting, smelting, or chemical or physical extraction)
H20   Solvent recovery (including distillation, evaporation, fractionation or extraction)
H39   Other recovery or reclamation for reuse (including acid regeneration or other chemical reaction process)

Section 8.10. Source Reduction Activity Codes
Good Operating Practices

W13   Improved maintenance scheduling, record keeping, or procedures
W14   Changed production schedule to minimize equipment and feedstock changeovers
W15   Introduced in-line product quality monitoring or other process analysis system
W19   Other changes in operating practices

Inventory Control

W21   Instituted procedures to ensure that materials do not stay in inventory beyond shelf-life
W22   Began to test outdated material - continue to use if still effective
W23   Eliminated shelf-life requirements for stable materials
W24   Instituted better labeling procedures
W25   Instituted clearinghouse to exchange materials that would otherwise be discarded
W29   Other changes in inventory control

Spill and Leak Prevention

W31   Improved storage or stacking procedures
W32   Improved procedures for loading, unloading, and transfer operations
W33   Installed overflow alarms or automatic shut-off valves
W35   Installed vapor recovery systems
W36   Implemented inspection or monitoring program of potential spill or leak sources
W39   Other changes made in spill and leak prevention

Raw Material Modifications

W41   Increased purity of raw materials
W42   Substituted raw materials
W43   Substituted a feedstock or reagent chemical with a different chemical
W49   Other raw material modifications made

Process Modifications

W50   Optimized reaction conditions or otherwise increased efficiency of synthesis
W51   Instituted recirculation within a process
W52   Modified equipment, layout, or piping
W53   Use of a different process catalyst
W54   Instituted better controls on operating bulk containers to minimize discarding of empty containers
W55   Changed from small volume containers to bulk containers to minimize discarding of empty containers
W56   Reduced or eliminated use of an organic solvent
W57   Used biotechnology in manufacturing process
W58   Other process modifications

Cleaning and Degreasing

W59   Modified stripping/cleaning equipment
W60   Changed to mechanical stripping/cleaning devices (from solvents or other materials)
W61   Changed to aqueous cleaners (from solvents or other materials)
W63   Modified containment procedures for cleaning units
W64   Improved draining procedures
W65   Redesigned parts racks to reduce drag out
W66   Modified or installed rinse systems
W67   Improved rinse equipment design
W68   Improved rinse equipment operation
W71   Other cleaning and degreasing modifications

Surface Preparation and Finishing

W72   Modified spray systems or equipment
W73   Substituted coating materials used
W74   Improved application techniques
W75   Changed from spray to other system
W78   Other surface preparation and finishing modifications

Product Modifications

W81   Changed product specifications
W82   Modified design or composition of products
W83   Modified packaging
W84   Developed a new chemical product to replace a previous chemical product
W89   Other product modifications

Section 8.10. Methods Used to Identify Source Reduction Activities

For each source reduction activity, enter up to three of the following codes that correspond to the method(s) which contributed most to the decision to implement that activity.

T01   Internal Pollution Prevention Opportunity Audit(s)
T02   External Pollution Prevention Opportunity Audit(s)

T03	Materials Balance Audits	T08	Federal Government Technical Assistance Program
T04	Participative Team Management		
T05	Employee Recommendation (independent of a formal company program)	T09	Trade Association/Industry Technical Assistance Program
T06	Employee Recommendation (under a formal company program)	T10	Vendor Assistance
		T11	Other
T07	State Government Technical Assistance Program		

B.2 Reporting the Waste Management of Metals

This appendix outlines how the TRI-MEweb reporting software restricts reporting for metals when the specific data element or waste management code is not applicable for a particular chemical. Below is a list of metals divided into four groups along with charts that help explain where quantities of these chemicals can and cannot be reported on the Form R using TRI-MEweb. In addition, there are charts that explain restrictions on reporting waste management codes for the toxic chemicals in each of the four groups. This appendix only shows where reporting is restricted in TRI-MEweb, it does not indicate every situation where a metal should not be reported in a specific section of the form. For example, TRI-MEweb does not restrict the reporting of most individually-listed metal compounds as used for energy recovery (Sections 8.2 and 8.3) even though some of these chemicals do not have a heat value greater than 5000 British thermal units (Btu) and, thus, cannot be combusted for energy recovery. It is left to the facility to decide which of these toxic chemicals can be used for energy recovery. If you are not using TRI-MEweb this appendix can serve as a guide to help you understand where it is not appropriate to report certain quantities of toxic chemicals or waste management codes on your Form R.

Parent Metals:

Antimony
Arsenic
Barium
Beryllium
Cadmium
Chromium
Cobalt
Copper
Lead
Manganese
Mercury
Nickel
Selenium
Silver
Thallium

Metal Compound Categories:

Antimony Compounds
Arsenic Compounds
Barium Compounds
Beryllium Compounds
Cadmium Compounds
Chromium Compounds
Cobalt Compounds
Copper Compounds
Lead Compounds
Manganese Compounds
Mercury Compounds
Nickel Compounds
Selenium Compounds
Silver Compounds
Thallium Compounds
Vanadium Compounds
Zinc Compounds

Metals with Qualifiers:

Aluminum (fume or dust)
Vanadium (except when in an alloy)
Zinc (fume or dust)

Individually-Listed Metal Compounds:

Bis(tributylin) oxide
Triphenyltin hydroxide
Triphenyltin chloride
Molybdenum trioxide
Thorium dioxide
Asbestos (friable)
Aluminum oxide (fibrous forms)
Tributyltin fluoride

Tributyltin methacrylate
Titanium tetrachloride
Boron trifluoride
Metiram
Boron trichloride
Zineb
Maneb
Fenbutatin oxide
Iron pentacarbonyl
Ferbam
C.I. Direct Brown 95
Osmium tetroxide
Aluminum phosphide
C.I. Direct Blue 218

Sections 5.3 - Discharges to Water and 6.1 - Transfers to POTWs

The following chart indicates which metals can be reported as released to water in Section 5.3 or to POTW's in Section 6.1. Only zinc (fume or dust) and aluminum (fume or dust) are not reported in these sections because the fume or dust form of a toxic chemical cannot exist in water.

Form R Section in Part II	Parent Metals	Metal Category Compounds	Metals with Qualifiers	Individually-listed Metal Compounds
Section 5.3 - Discharges to receiving streams or water bodies	All	All	Vanadium (except when contained in an alloy)	All except Asbestos
Section 6.1- Discharges to POTWs	All	All	Vanadium (except when contained in an alloy)	All except Asbestos

Section 6.2. Transfers to Other Off-Site Locations

Any toxic chemical may be reported in Section 6.2. However, TRI-MEweb will not allow certain M codes to be used when reporting metals. The chart below indicates which M codes can be reported in Section 6.2 for the four groups of metals. Note that all disposal M codes other than M41 and M62 can be used for all toxic chemicals. Code M24 is only made available for the four groups of metals.

Waste Management Code for Section 6.2	Parent Metals	Metal Category Compounds	Metals with Qualifiers	Individually-listed Metal Compounds
M41 and M62 (disposal codes-for metals only)	All	All	Vanadium (except when contained in an alloy)	All except Asbestos
M56 and M92 (energy recovery codes)	None	None	None	All except Asbestos[1]
M20 and M28 (recycling codes)	None	None	None	All
M24, M26 and M93 (recycling codes)	All	All	All	All
M40, M50, M54, (treatment codes)	None	None	All except Vanadium (except when contained in an alloy)	All
M61, M69, M95 (treatment codes)	Barium[2]	Barium Compounds[2]	Same as above	All

Section 7A. On-site Waste Treatment Methods and Efficiency

TRI-MEweb allows any toxic chemical to be reported in Section 7A, however, it limits reporting in two ways. First, TRI-MEweb limits the treatment codes that can be reported based on the General Waste Stream Code selected. If a TRI-MEweb user selects General Waste Stream code "A – Gaseous", all Waste Treatment Codes are made available. However, if a user selects from the remaining three General Waste Stream Codes (W - Wastewater, L - Liquid waste streams, or S - Solid waste streams), the "Air Emissions Treatment" Waste Treatment Codes are not made available. Second, the software restricts reporting for certain toxic chemicals with qualifiers. When reporting zinc (fume or dust) or aluminum (fume or dust) TRI-MEweb will not allow the user to select General Waste Stream Codes W-Wastewater and L-Liquid waste streams because the fume or dust form of a toxic chemical cannot exist in a liquid or water waste. For asbestos (friable) only S - Solid or A - Gaseous can be selected. When reporting hydrochloric acid (acid aerosols) or sulfuric acid (acid aerosols) only A - Gaseous can be selected.

Crosswalk for Section 7A, Column B. Waste Treatment Method (s) Sequence

Air Emissions Treatment (applicable to gaseous waste streams only) (No change — same as previous codes)			
A01	Flare		
A02	Condenser		
A03	Scrubber		
A04	Absorber		
A05	Electrostatic Precipitator		
A06	Mechanical Separation		
A07	Other Air Emission Treatment		

Biological Treatment:			
Previous Codes		**New Codes (adapted from RCRA Hazardous Waste Management Codes)**	
B11	Aerobic	H081	Biological treatment with or without precipitation
B21	Anaerobic	H081	Biological treatment with or without precipitation
B31	Facultative	H081	Biological treatment with or without precipitation
B99	Other Biological Treatment	H081	Biological treatment with or without precipitation

Chemical Treatment:			
Previous Codes		**New Codes (adapted from RCRA Hazardous Waste Management Codes)**	
C01	Chemical Precipitation B Lime or Sodium Hydroxide	H071	Chemical reduction with or without precipitation
C02	Chemical Precipitation B Sulfide	H071	Chemical reduction with or without precipitation
C09	Chemical Precipitation B Other	H077	Other chemical precipitation with or without pre-treatment
C11	Neutralization	H121	Neutralization
C21	Chromium Reduction	H071	Chemical reduction with or without precipitation
C31	Complexed Metals Treatment (other than pH adjustment)	H129	Other treatment
C41	Cyanide Oxidation B Alkaline Chlorination	H073	Cyanide destruction with or without precipitation
C42	Cyanide Oxidation B Electrochemical	H073	Cyanide destruction with or without precipitation
C43	Cyanide Oxidation B Other	H073	Cyanide destruction with or without precipitation
C44	General Oxidation (including Disinfection) B Chlorination	H075	Chemical oxidation
C45	General Oxidation (including Disinfection) B Ozonation	H075	Chemical oxidation
C46	General Oxidation (including Disinfection) B Other	H075	Chemical oxidation
C99	Other Chemical Treatment	H129	Other treatment

Chemical Treatment:			
Previous Codes		**New Codes (adapted from RCRA Hazardous Waste Management Codes)**	
Incineration/Thermal Treatment: (Note: Only report combustion for the purposes of incineration/thermal treatment in Section 7A. If the method involves combustion for the purposes of energy recover, report as U01, U02, or U03 in Section 7B. If the method involves combustion for the purposes of materials recovery, report as H39 in Section 7C.)			
F01	Liquid Injection	H040	Incineration B thermal destruction other than use as a fuel
F11	Rotary Kiln with Liquid Injection Unit	H040	Incineration B thermal destruction other than use as a fuel
F19	Other Rotary Kiln	H040	Incineration B thermal destruction other than use as a fuel
F31	Two Stage	H040	Incineration B thermal destruction other than use as a fuel
F41	Fixed Hearth	H040	Incineration B thermal destruction other than use as a fuel
F42	Multiple Hearth	H040	Incineration B thermal destruction other than use as a fuel
F51	Fluidized Bed	H040	Incineration B thermal destruction other than use as a fuel
F61	Infra-Red	H040	Incineration B thermal destruction other than use as a fuel
F71	Fume/Vapor	H040	Incineration B thermal destruction other than use as a fuel
F81	Pyrolytic destructor	H040	Incineration B thermal destruction other than use as a fuel
F82	Wet air oxidation	H076	Wet air oxidation
F83	Thermal Drying/Dewatering	H122	Evaporation
F99	Other Incineration/Thermal Treatment	H040	Incineration B thermal destruction other than use as a fuel

Physical Treatment:			
Previous Codes		**New Codes (adapted from RCRA Hazardous Waste Management Codes)**	
P01	Equalization	H129	Other treatment
P09	Other blending	H129	other treatment
P11	Settling/clarification	H123	Settling or clarification
P12	Filtration	H123	Settling or clarification
P13	Sludge dewatering (non-thermal)	H101	Sludge treatment and/or dewatering
P14	Air flotation	H124	Phase separation
P15	Oil skimming	H124	Phase separation
P16	Emulsion breaking B thermal	H124	Phase separation
P17	Emulsion breaking B chemical	H124	Phase separation
P18	Emulsion breaking B other	H124	Phase separation

Physical Treatment:			
Previous Codes		**New Codes (adapted from RCRA Hazardous Waste Management Codes)**	
P19	Other liquid phase separation	H124	Phase separation
P21	Adsorption B Carbon	H082	Adsorption
P22	Adsorption B Ion exchange (other than for recovery/reuse)	H082	Adsorption
P23	Adsorption B Resin	H082	Adsorption
P29	Adsorption B Other	H082	Adsorption
P31	Reverse Osmosis (other than for recover/reuse)	H129	Other treatment
P41	Stripping B Air	H083	Air or steam stripping
P42	Stripping B Steam	H083	Air or steam stripping
P49	Stripping B Other	H083	Air or steam stripping
P51	Acid Leaching (other than for recovery/reuse)	H129	Other treatment
P61	Solvent Extraction (other than recovery/reuse)	H129	Other treatment
P99	Other Physical Treatment	H129	Other treatment

Solidification/Stabilization:			
Previous Codes		**New Codes (adapted from RCRA Hazardous Waste Management Codes)**	
G01	Cement processes (including silicates)	H111	Stabilization or chemical fixation prior to disposal
G09	Other Pozzolonic Processes (including silicates)	H111	Stabilization or chemical fixation prior to disposal
G11	Asphaltic Techniques	H111	Stabilization or chemical fixation prior to disposal
G20	Thermoplastic Techniques	H111	Stabilization or chemical fixation prior to disposal
G99	Other Solidification Processes	H111	Stabilization or chemical fixation prior to disposal

Section 7B. On-site Energy Recovery Processes

The chart below indicates which energy recovery codes can be reported in TRI-MEweb in Section 7B for the four groups of metals.

Energy Recovery Code for Section 7B	Parent Metals	Metal Category Compounds	Metals with Qualifiers	Individually-listed Metal Compounds
U01, U02, U03	None	None	None	All except Asbestos[1]

Section 7C. On-site Recycling Processes

Any chemical can be reported in Section 7C. However, certain waste management codes should not be reported for certain toxic chemicals. The chart below indicates which codes can be reported in Section 7C when using TRI-MEweb.

Recycling Code for Section 7C	Parent Metals	Metal Category Compounds	Metals with Qualifiers	Individually-listed Metal Compounds
H10 (this code is for metals only)	All	All	All	All
H20	None	None	None	All
H39	All	All	All	All

Crosswalk for Section 7C. On-site Recycling Processes

Previous Codes		New Codes (adapted from RCRA Hazardous Waste Management Codes)	
R11	Solvents/Organics Recovery B Batch Still Distillation	H20	Solvent Recovery (including distillation, evaporation, fractionation or extraction)
R12	Solvents/Organics Recovery B Thin-Film Evaporation	H20	Solvent Recovery (including distillation, evaporation, fractionation or extraction)
R13	Solvents/Organics Recovery B Fractionation	H20	Solvent Recovery (including distillation, evaporation, fractionation or extraction)
R14	Solvents/Organics Recovery B Solvent Extraction	H20	Solvent Recovery (including distillation, evaporation, fractionation or extraction)
R19	Solvents/Organics Recovery B Other	H20	Solvent Recovery (including distillation, evaporation, fractionation or extraction)
R21	Metals Recovery B Electrolytic	H10	Metal Recovery (by retorting, smelting, or chemical or physical extraction)
R22	Metals Recovery B Ion Exchange	H10	Metal Recovery (by retorting, smelting, or chemical or physical extraction)
R23	Metals Recovery B Acid Leaching	H10	Metal Recovery (by retorting, smelting, or chemical or physical extraction)
R24	Metals Recovery B Reverse Osmosis	H10	Metal Recovery (by retorting, smelting, or chemical or physical extraction)
R26	Metals Recovery B Solvent Extraction	H10	Metal Recovery (by retorting, smelting, or chemical or physical extraction)
R27	Metals Recovery B High Temperature	H10	Metal Recovery (by retorting, smelting, or chemical or physical extraction)

Previous Codes		New Codes (adapted from RCRA Hazardous Waste Management Codes)	
R28	Metals Recovery B Retorting	H10	Metal Recovery (by retorting, smelting, or chemical or physical extraction)
R29	Metals Recovery B Secondary Smelting	H10	Metal Recovery (by retorting, smelting, or chemical or physical extraction)
R30	Metals Recovery B Other	H10	Metal Recovery (by retorting, smelting, or chemical or physical extraction)
R40	Acid Regeneration	H39	Other recovery or reclamation for reuse (including acid regeneration or other chemical reaction process)
R99	Other Reuse or Recovery	H39	Other recovery or reclamation for reuse (including acid regeneration or other chemical reaction process)

Section 8. Source Reduction and Recycling Activities

The chart below indicates which metals can be reported in Sections 8.2, 8.3, 8.6 and 8.7 of the Form R when using *TRI-MEweb*. Note that all toxic chemicals can be reported in Sections 8.1, 8.4, 8.5 and 8.8.

Waste Management Activity	Parent Metals	Metal Category Compounds	Metals with Qualifiers	Individually-listed Metal Compounds
Quantity used for energy recovery on site and off site (Sections 8.2 and 8.3)	None	None	None	All except Asbestos[2]
Quantity treated for destruction on site and off site (Sections 8.6 and 8.7)	None except Barium[2]	None except Barium Compounds[2]	All except Vanadium (except when contained in an alloy)	All

---

[1] Although TRI-MEweb does not restrict reporting of most individually-listed metal compounds as transferred off site for energy recovery, only chemicals with a heat value greater than 5000 British thermal units that are combusted in a device that is an industrial furnace or boiler (40 CFR Section 372.3) should be reported as used for energy recovery.

[2] The toxic chemical category barium compounds (N040) does not include barium sulfate. Because barium sulfate is not a listed toxic chemical, the conversion in a waste stream of barium or barium compound to barium sulfate is considered treatment for destruction (40 CFR Section 372.3).

Appendix C. Electronic Facility Data Profiles and Common Errors in Completing Form R Reports and Form A Certification Statements

EPA wishes to ensure that facilities submit all required TRI chemical submissions in a timely manner so that the information may be included in its national database, annual public data release, now known as the TRI National Analysis and other information products. Moreover, EPA seeks to ensure that all submitted data are complete and accurate. This appendix provides an overview of the Electronic Facility Data Profile (eFDP), an important document that EPA uses as a receipt to our reporting facilities to ensure consistent, complete, and accurate submissions. This appendix also provides specific guidance to avoid common errors in completing Form Rs and Form A Certification Statements, including errors in threshold determination, misapplication of exemptions, and activities involving a reportable chemical, any of which may result in the erroneous non-reporting of a chemical. Most errors in this appendix often occur on paper form submissions. Using TRI-MEweb to electronically submit your TRI forms can help avoid errors because the software checks for errors before submission. If errors are identified in your eFDP, the preferred method to revise your TRI form is by the use of the TRI-MEweb application via EPA's Central Data Exchange (CDX).

C.1 Electronic Facility Data Profile (eFDP)

The eFDP report are made available via TRI-MEweb to a reporting facility in response to any submission the TRI Data Processing Center receives either through a paper form or an electronic submission that has completed processing into EPA database. If the Technical contact, preparer or certifying official provided an email address in the Form R/Form A, they will receive a real-time email notifying them when their eFDP has been updated. The email will contain information explaining how to create a CDX user account and how to add the TRI-MEweb application. Beginning in April 2012, all reporting facilities that reported electronically using TRI-MEweb or mailed a hard-copy TRI form to EPA's Data Processing Center (DPC), will obtain and review their eFDP report within the TRI-MEweb application. Any reporting facility's official may confirm and review their submitted TRI data to EPA by viewing their electronic Facility Data Profile (eFDP) on the Internet by logging into their CDX account and clicking the *TRI-MEweb: TRI Made Easy Web* link from their MyCDX page. This will open the "Welcome" page of the TRI-MEweb application. On the "Welcome" page, they can follow the instructions for viewing the eFDP. It is very important that you review your eFDP report because EPA may have transcribed your data incorrectly into the database from paper submissions or your reporting facility may have

incorrectly entered an incorrect waste quantity in your TRI-MEweb submission, or your reporting facility incorrectly listed the chemical category. Reviewing the eFDP allows reporting facilities to conduct final checks of the data submitted to EPA before it is released to the public. If you have questions regarding your eFDP, please send an email to helpdesk@epacdx.net or call 1 (888) 890-1955.

An eFDP report is comprised of the following sections:

Facility Information. This section displays all facility-specific data, including TRI Facility Identification (TRIFID), facility name, facility address, facility mailing address, North American Industry Classification System code (NAICS), and other facility data. Errors related to facility information will be marked in this section.

Instructions Page. This page provides instructions on how to review and respond to the eFDP.

Certification Statement Signature Page. This page provides the Certification Statement to be signed by a facility owner/operator or senior management official if using the eFDP to make a revision.

Chemical Report Summary. This section lists all chemicals reported by the facility for each reporting year covered by the eFDP. For example, if the eFDP is responding to five original chemical submissions for Reporting Year 2012 and revisions to one chemical for Reporting Year 2011, a list of all chemicals for both years will appear.

Errors/Alerts Identified In This Report: Non-Technical Data Changes (NDC), Notices of Technical Errors (NOTE), Notices of Significant Error (NOSE), and Data Quality Alerts (DQA). eFDPs identify three different types of errors: NDCs, NOTEs and NOSEs and one type of alert called Data Quality Alert (DQA). See explanations in section B.

Error Summary Page. The Error Summary Page provides facilities an error/alert count for each chemical submission.

Chemical Reports. All recently submitted and processed Form R or Form A data (i.e., chemical specific data) are displayed in the chemical reports under the appropriate facility or subordinate facility names. The eFDP report displays facsimiles for chemical reports for submissions received during the current calendar year and revisions or responses to eFDPs only. For example, if a facility originally reported five chemicals for Reporting Year 2012, and subsequently revises only one chemical submission, the facility will receive an eFDP for Reporting Year 2012 with only the revised chemical included in the Chemical Reports section. Hence there may be fewer chemical reports than chemicals listed in the Chemical Summary section. If only facility level changes have occurred (i.e., Part I of the Form R or A), this section is not provided.

Data Quality Alerts. TRI provides Data Quality Alerts (DQA) in eFDP reports. The DQA informs facilities of possible reporting errors by flagging data trends that are outside the norm. For example, if a facility reports a change in the release of a chemical that is over 25% compared to last year, a DQA will be triggered. This is offered to assist facilities in ensuring accurate reporting.

C.2 Levels of Errors Identified in eFDPs: Notice of Non-Technical Data Change (NDC), Notice of Technical Errors (NOTE), Notice of Significant Errors (NOSE), Notice of Noncompliance (NON)

eFDP Error Reporting. In addition to echoing back the information a facility has submitted, eFDPs are used to identify potential errors and provide Data Quality Alerts. Errors are more likely to be submitted on paper forms because forms submitted through TRI-MEweb benefit from built-in software to detect some errors during data entry, prompting the user to correct them before the forms are transmitted, certified, and submitted to EPA. Errors are still possible on forms submitted through TRI-MEweb and this appendix will indicate whether specific errors can occur on paper forms or TRI-MEweb submissions or both.

As submission information is entered into EPA's national database, a series of automated data quality checks are performed. Some error messages will indicate where the TRI Data Processing Center has made minor clerical changes to submissions. The data quality checks are useful to identify potential errors with certain data fields such as TRI Facility Identification, facility name, county spelling, as well as to perform validation checks to ensure consistency among data elements within a given Form R or Form A. These data quality checks, however, cannot detect whether release, transfer, or waste management quantities were calculated or entered accurately.

Within an eFDP report, there may be up to three different types of errors identified.

Non-Technical Data Change (NDC)

Applies to: Paper forms only

A **Non-Technical Data Change** (NDC) notifies you of simple, clerical errors that the TRI Data Processing Center has corrected for you. It is not necessary to respond to a NDC. The TRI Data Processing Center will correct simple, clerical errors that are not technical or scientific - a "**non-technical data change**." For example, if a facility transposes CAS numbers (e.g., the submitter lists 7623-00-0 for sodium nitrite instead of 7632-00-0), the TRI Data Processing Center will correct this clerical error and display the correct information on the facility's eFDP. If a facility lists a specific glycol ethers subcategory, the TRI Data Processing Center will replace this subcategory with the reportable name "certain glycol ethers." The messages used on eFDPs to report non-technical data changes are shown at the end of this appendix under the heading "C.5 Messages Used to Report Notices of Technical Errors (NOTEs) and Non-technical Data Changes (NDCs)." This type of error is flagged for correction during data entry when using TRI-MEweb and needs to be addressed by the facility before submission is submitted and processed by EPA. Therefore, NDC's are not possible in a TRI-MEweb submission.

Notice of Technical Error (NOTE)

Applies to: Paper forms only

A **Notice of Technical Error** (NOTE) highlights inconsistencies or miscalculations that may distort your facility's information in EPA's public data products or skew analyses. Incomplete addresses, no technical or public contact provided, missing or invalid NAICS codes, or the use of range codes to report PBT chemical releases are all examples of technical errors. You should respond to NOTEs as soon as possible. These types of errors could be corrected by the reporting facility on its eFDP (or provide the TRI Data Processing Center with a brief explanation why they do not believe that it is an error) or submit a revised Form R or Form A. Depending upon when your changes are received, there may or may not be sufficient time to incorporate them into EPA's database before your report has been released to the public Technical errors do not prevent submissions from being entered into the data management system, but indicate inconsistencies or miscalculations in the submitted form. These errors can distort public information products and skew any analyses if not corrected. The messages used on eFDPs to report NOTEs are shown below at the end of this appendix under the heading "C.5 Messages Used to Report Notices of Technical Errors (NOTEs) and Non-technical Data Changes (NDCs)." This type of error is flagged for correction during data entry when using TRI-MEweb and need to be addressed by facility before submission is submitted and processed by EPA. Therefore, NOTE's are not possible in a TRI-MEweb submission.

Notices of Significant Errors (NOSE)

Applies to: Paper forms and TRI-MEweb submissions

The most serious errors are classified as **Notices of Significant Errors (NOSE)**. The eFDP contains the Notice of Significant Error if applicable. Significant errors prevent submissions from being entered into the TRI Data Processing Center data management system or do not allow the TRI Data Processing Center to verify the authenticity of the submission. Invalid forms, missing pages, no chemical name or CAS number are examples of significant errors. These types of errors could be corrected by the reporting facility on their eFDP, or the reporting facility could submit a revised Form R or Form A, or the reporting facility could provide the TRI Data Processing Center with a brief explanation why they do not believe that it is an error. A facility must respond to a Notice of Significant Error within 21 days of receipt. Failure to respond within the initial 21 day requirement could result in the issuance of a Notice of Noncompliance (NON). A Notice of Noncompliance is not included in an eFDP and is mailed separately.

Reporters will receive a NOSE for failure to certify a submission (i.e. not signing paper forms). This includes any electronic submission that is not certified in the TRI-MEweb system as of July 1st, 2012 for which the user has not submitted certification via another reporting media, such as paper.

Notice of Noncompliance (NON)

Applies to: Paper forms and TRI-MEweb submissions

The Agency will issue a **Notice of Noncompliance (NON)** to a facility for failure to respond to a Notice of Significant Error (NOSE) within the required period. A NON suggests that a facility should take corrective action within 30 days and respond to the Agency that corrective action has been taken. If a facility fails to respond to the NON within the required time period, the Agency may take further action.

Record Keeping

Facilities must keep copies, for three years, of submitted Form R reports and Form A certifications and all documentation used to complete their submissions in accordance with 40 CFR 372.10. This documentation should include calculations for threshold determinations, the basis of exemptions applied, and the estimation techniques and data used for all quantities reported on the Form R and Form A. TRI-MEweb stores several years worth (7 years in RY 2012) of submitted chemical release data that can be accessed to be printed for your records.

C.3 Common Errors in Completing Form R Reports and Form A Certification Statements.

The following section lists the most common errors that reporting facilities have encountered when submitting paper or TRI-MEweb submissions to EPA. Some of these errors are not detected nor listed on an eFDP report. Errors that are not detectable are hard to evaluate by EPA because they could be valid submissions and can only be determined to be incorrect by the reporting facility. Reporting facilities should review their submission to ensure these common errors are not present in their forms before submitting them to EPA.

General Considerations

Applies to: Paper forms only

Lack of signed certification statement. If you choose not to send your TRI submissions via the paperless CDX process, you must sign and submit Part I, Section 3 of your hard copy submission. Although EPA accepts paper submissions, EPA strongly encourages you to send your submission via TRI-MEweb and CDX. This error type is listed on an eFDP as a NOSE.

Incomplete Forms. A complete Form R report for a single EPCRA section 313 chemical or single EPCRA section 313 chemical category consists of six pages stapled together. By using TRI-MEweb and CDX, errors such as this would not occur. Each chemical submission must have its own page one. EPA cannot enter into the database data from a package that contains only one page 1, but several page 2s, 3s, 4s, 5s and/or 6s. Such forms are considered incomplete submissions. This error type is listed on an eFDP as a NOSE.

Threshold Determinations

Applies to: Paper forms and TRI-MEweb submissions

Calculating threshold determinations. Annual quantities manufactured, processed, or otherwise used for section 313 chemicals must be calculated, not surmised. The assumption that thresholds are exceeded commonly leads to error. This error type is not detected nor listed on an eFDP report.

Misclassification of EPCRA section 313 chemical activity. Failure to correctly classify an EPCRA section 313 chemical activity may result in an incorrect threshold determination. As a result, a facility may fail to submit the required Form R. This error type is not detected nor listed on an eFDP report.

EPCRA section 313 chemical activity overlooked. Many facilities believe that because the section 313 reporting requirement pertains to manufacturers, only the use of EPCRA section 313 chemicals in manufacturing processes must be examined. *Any activity* involving the manufacture, process, or otherwise use of an EPCRA section 313 chemical or chemical category must be included in threshold determinations. Commonly overlooked activities include importation of chemicals, generation of waste byproducts, processing of naturally occurring metals and metal category compounds in ore, manufacturing and processing intermediates, the use of chemicals for cleaning of equipment, and the generation of byproducts during combustion of coal and/or oil. Facilities should take a systematic approach to identify all chemicals and mixtures used in production and non-production capacities, including catalysts, well treatment chemicals, and wastewater treatment chemicals. This error type is not detected nor listed on an eFDP report.

Considering EPCRA section 313 chemicals in mixtures and other trade name products. EPCRA section 313 chemicals contained in mixtures (including ores and stainless steel alloys) and other trade name products must be factored into threshold determinations and release and other waste management determinations, provided that the *de minimis* exemption cannot be taken. When the EPCRA section 313 chemical being reported is a component in a mixture or other trade name product, report only the weight of the EPCRA section 313 chemical in the mixture. Refer to Section B.4f of this document to calculate the weight of an EPCRA section 313 chemical in a mixture or other trade name product. This error type is not detected nor listed on an eFDP report.

Overlooking manufacturing. Coincidental manufacturing must not be overlooked. If coal and/or fuel oil and other raw materials that contain EPCRA section 313 chemicals are used in boilers/burners, there is a potential for the coincidental manufacture of EPCRA section 313 chemicals such as sulfuric acid (acid aerosols), hydrochloric acid (acid aerosols), hydrogen fluoride, and metal category compounds. Additionally, manufacturing of EPCRA section 313 chemicals during waste treatment is commonly overlooked. For example, the treatment of nitric acid may result in the manufacturing of a reportable chemical (nitrate compounds). This error type is not detected nor listed on an eFDP report.

Container Residue

Overlooking container residue. Container residue must not be disregarded in release and other waste management calculations. Even a "RCRA empty" drum is expected to contain a residue and it must be considered for TRI reporting. Additionally, on-site drum rinsing and disposal of the rinsate will result in a release and other waste management activity. Refer to Estimating Releases and Waste Treatment Efficiencies for Toxic Chemical Reporting Forms. This error type is not detected nor listed on an eFDP report.

Part I. Facility Identification Information

Section 1. Reporting Year

* **Invalid Paper forms:** Hard copy submissions may be submitted using the TRI Form R and/or Form A Certification Statement applicable for that particular reporting year. EPA provides printable TRI forms from RY 2003 through RY 2012 on the TRI website at http://www.epa.gov/tri/reporting_materials/forms/. For reporters submitting RY 2011 and RY 2012 hard-copy forms, EPA recommends entering data using the electronically fillable fields in the RY 2011 and RY 2012 forms. RY 2010 and prior year forms are not electronically fillable and must be completed by hand or typewriter. You can also request older reporting forms under the *Contact Us* link on the TRI web site for TRI forms prior to RY 2003. Please sign and date the certification statement on Page 1 prior to mailing your TRI form(s) to EPA's DPC. This error type is listed on an eFDP as a NOSE.

* **Invalid TRI-MEweb Forms:** Users that prepare TRI forms using TRI-MEweb must pick the reporting year before starting to enter any chemical release data. Users may start a blank form or choose to import prior year data into current year forms from the *Form Summary Table* on the TRI-MEweb Welcome page after clicking on the (+) sign next to TRIFID of the reporting facility. If the preparer transmitted, certified and submitted a form with an incorrect reporting year selected, a revision of this form cannot change the reporting year field. Instead, the incorrect reporting year form must be withdrawn and resubmitted under the correct reporting year. This error type is not detected nor listed on an eFDP report.

Section 2. Trade Secret Information

Applies to: Paper forms only

Incorrect completion of trade secret information. The responses to trade secret questions in Part I Section 2 and Part II Section 1.3 of Form R/Form

A must be consistent. If trade secrecy is indicated, a sanitized Form R/Form A and two trade secret substantiations (one sanitized) must be submitted in the same package as the unsanitized trade secret Form R/Form A. Part II Section 1.3 should be blank if no trade secret claim is being made. Also, if you indicate in Part I, Section 2.1 that you are **not** claiming trade secret information, leave Part I, 2.2 blank. This error type is listed on an eFDP as a NOSE.

Section 3. Certification

Applies to: Paper forms only

Missing certification signature. If you are submitting your Form R and/or Form A by hardcopy, an original certification signature must appear on page 1 of every Form R and/or Form A Certification Statement submitted to EPA. Missing signatures will prevent forms from being processed by EPA. This error type is listed on an eFDP as a NOSE.

Applies to: TRI-MEweb submissions only

Uncertified TRI-MEweb submissions If you are submitting your Form R and/or Form A via TRI-MEweb and CDX, you must electronically sign the submission before it can be loaded into the TRI database, Uncertified electronic submissions will not be accepted and result in a Notice of Non-Compliance (NON) if they remain uncertified after the July 1st reporting deadline and you have not reported by another means. If your eFDP report is not created in TRI-MEweb after certifying your electronic submission, this may indicate that the certification process was not completed correctly. This error type is listed on an eFDP as a NOSE.

Section 4. Facility Identification

Questionable entries. Incorrect entries may be corrected by the reporting facility though a revision. The use of the TRI-MEweb software may prevent such errors from occurring. Questionable entries may include:

Applies to: Paper forms and TRI-MEweb submissions

- Incorrect street address;
- Incorrect ZIP codes;
- Invalid County names;
- Invalid NAICS codes;
- Invalid Dun & Bradstreet numbers;

Note: These error types are not detected nor listed on an eFDP report.

Applies to: Paper forms only

- Missing street address;
- Missing ZIP codes

- Missing County Names

Note: These error types are listed on an eFDP as a NOSE.

Part II. Chemical-Specific Information

Section 1. Toxic Chemical Identity

Applies to: Paper forms only

Reporting chemical abstract service (CAS) registry numbers in Section 1.1. In 1992, EPA assigned alphanumeric category codes to the twenty chemical categories for the purposes of reporting the CAS number field in Section 1.1. Incorrect use of chemical category codes have caused errors on TRI forms requiring forms to be withdrawn and re-submitted. When completing a Form R for a chemical category, the appropriate code for that category must be provided in Section 1.1. The CAS numbers are listed in Table II: "Section 313 Toxic Chemical List," and if needed, the category codes are listed in Appendix B: "Reporting Codes for EPA Form R." Category guidance documents are listed in the Chemical and Industry Guidance Documents section in this document. This error type is not detected nor listed on an eFDP report.

Invalid chemical identification in Section 1.2. The CAS number and the chemical name reported here must exactly match the listed official EPCRA section 313 CAS number and EPCRA section 313 chemical name. This error type is listed on an eFDP as a NOTE.

Applies to: Paper forms and TRI-MEweb submissions

Failure to check for synonyms. Some reportable chemicals (especially glycol ethers and toluene diisocyanates) have many synonyms that do not readily imply they are in the category. For example, benzene,1,3-diisocyanatomethyl may not be readily recognized as toluene diisocyanate (mixed isomers). This error type is not detected nor listed on an eFDP report.

Generic chemical name used in Section 1.3. A generic chemical name should only be provided if the section 313 chemical identity is claimed as a trade secret. Generic names should not be used if no trade secret submissions are being claimed by a reporting facility. This error type is listed on an eFDP as a NOSE.

Failure to consider an EPCRA section 313 chemical qualifier. Only EPCRA section 313 chemicals in the form specified in the qualifier require reporting under section 313 and should be reported on Form R with the appropriate qualifier in parentheses. For example, isopropyl alcohol is listed on the EPCRA section 313 chemical list with the qualifier manufacturing- strong acid process, no supplier notification. Thus, the ONLY facilities that should report this EPCRA section 313 chemical are those that manufacture isopropyl alcohol by the strong acid process. This error type is not detected nor listed on an eFDP report.

Section 2. Mixture Component Identity

Applies to: Paper forms and TRI-MEweb submissions

Identifying chemicals used in mixtures. Facilities should carefully review the most recent MSDS or supplier notification for every mixture brought on-site to identify all section 313 chemicals used during a reporting year. Although some mixtures may not have MSDSs, the best readily available information should be used to determine the presence of EPCRA section 313 chemicals in ores and alloys. This error type is not detected nor listed on an eFDP report.

Mixture names in Section 2.1. Mixture names are to be entered here only if the supplier is claiming the identity of the EPCRA Section 313 chemical a trade secret and that is the sole identification. Mixture names that include the name or CAS number of one or more EPCRA Section 313 chemicals are not valid uses of the mixture name field. This error type is not detected nor listed on an eFDP report.

Section 3. Activities and Uses of the Toxic Chemical at the Facility

Applies to: Paper forms and TRI-MEweb submissions

Reporting EPCRA section 313 chemical activity. EPCRA section 313 chemical activity is commonly overlooked or misclassified. *Any activity* involving the manufacture, process, or otherwise use of an EPCRA Section 313 chemical must be examined. For example, waste treatment operations otherwise use EPCRA Section 313 chemicals to treat waste streams and may coincidentally manufacture an additional EPCRA Section 313 chemical as a result of the treatment reaction. Such activity must be considered. Further, EPCRA Section 313 chemical activity must be correctly classified as either "manufactured," "processed," or "otherwise used."

Section 3.1 Manufacture means to produce, prepare, compound, or import an EPCRA Section 313 chemical.

Section 3.2 Process means the preparation of an EPCRA Section 313 chemical after its

manufacture, which usually includes the incorporation of the EPCRA Section 313 chemical into the final product, for distribution in commerce.

Section 3.3 Otherwise use encompasses any use of an EPCRA Section 313 chemical that does not fall under the terms "manufacture" or "process," and includes treatment for destruction, stabilization (without subsequent distribution in commerce), disposal, and other use of an EPCRA Section 313 chemical, including an EPCRA Section 313 chemical contained in a mixture or other trade name product. Otherwise use of an EPCRA Section 313 chemical does not include disposal, stabilization (without subsequent distribution in commerce), or treatment for destruction unless:

1. The EPCRA Section 313 chemical that was disposed of, stabilized, or treated for destruction was received from off-site for the purposes of further waste management; or

2. The EPCRA Section 313 chemical that was disposed of, stabilized, or treated for destruction was manufactured as a result of waste management activities on materials received from off-site for the purposes of further waste management activities.

For example, solvents in paint applied to a manufactured product are often misclassified as processed, instead of otherwise used. Because the solvents are not incorporated into the final product, the solvent is being otherwise used, not processed. This error type is not detected nor listed on an eFDP report.

Section 4. Maximum Amount of the Toxic Chemical On-site at Any Time During the Calendar Year

Applies to: Paper forms only

Maximum amount on-site left blank. Form has failed to provide the appropriate code for maximum amount on site. This error type is listed on an eFDP as a NOSE.

Incorrect units of measure. If amounts are reported in units other than pounds (e.g., metric units) or with exponential numbers, EPA may require a revision of the Form R/Form A submitted. The exception is for the reporting of dioxin and dioxin-like compounds where the amounts are reported in grams. This error type is not detected nor listed on an eFDP report.

Section 5. Quantity of the Toxic Chemical Entering Each Environmental Medium On-site

Applies to: Paper forms and TRI-MEweb submissions

Incorrectly reporting stack emissions. Fugitive emissions from general indoor air should not be reported as stack missions when released from a single building vent. Additionally, stack emissions from storage tanks, including loading, working, and breathing losses from tanks, should not be overlooked or reported as fugitive emissions. This error type is not detected nor listed on an eFDP report.

Overlooking releases to land. Section 313 chemicals placed in stockpiles or in surface impoundments should be reported as a "release to land" even if no Section 313 chemicals leak from these sources. Quantities of Section 313 chemicals land-treated should be reported as a release to land. This error type is not detected nor listed on an eFDP report.

Section 6. Transfers of the Toxic Chemical in Wastes to Off-site Locations

Applies to: Paper forms and TRI-MEweb submissions

Reporting discharges to POTWs in Section 6.1. When quantities of a listed mineral acid are neutralized to a pH of 6 or greater, the quantity reported as discharged to a POTW should be reported as zero. It is incorrect to enter "NA" (Not Applicable), in such a situation. This error type is not detected nor listed on an eFDP report.

Reporting other off-site transfers in Section 6.2. Any quantities reported in Sections 8.1, 8.3, 8.5, and 8.7 as sent off-site for disposal, treatment, energy recovery, or recycling, respectively, must also be reported in Section 6.2 along with the receiving location and appropriate off-site activity code. This error type is not detected nor listed on an eFDP report.

Section 7A. On-Site Waste Treatment Methods and Efficiency

Applies to: Paper forms and TRI-MEweb submissions

Failure to report waste treatment methods in Section 7A. Waste treatment methods used to treat waste streams containing EPCRA Section 313 chemicals, and the efficiencies of these methods, must be reported on Form R. Information must be entered for all waste streams, even if the waste treatment method does not affect the EPCRA Section 313 chemical. If no waste treatment is performed on waste streams containing the EPCRA Section 313 chemical, the box marked Not Applicable in Section 7A should be checked on Form R. This error type is not detected nor listed on an eFDP report.

Incorrect reporting of waste treatment methods in Section 7A. The type of waste stream, waste treatment efficiency, and waste treatment method for each waste stream are required to be reported on Form R using specific codes. The waste treatment codes are listed in Appendix B: Reporting Codes for EPA Form R. A table is also provided in Appendix B that displays a crosswalk between the old codes and new ones for reporting year 2005. This error type is not detected nor listed on an eFDP report.

Section 7B. On-Site Energy Recovery Processes

Applies to: Paper forms and TRI-MEweb submissions

Reporting on-site energy recovery methods in Section 7B. When a quantity is reported in Section 8.2 as combusted for energy recovery on-site, the type of energy recovery system used must be reported in Section 7B, and vice versa. This error type is not detected nor listed on an eFDP report.

Section 7C. On-Site Recycling Processes

Applies to: Paper forms and TRI-MEweb submissions

Reporting on-site recycling methods in Section 7C. When a quantity is reported in Section 8.4 as recycled on-site, the type of recovery method must be reported in Section 7C, and vice versa. This error type is not detected nor listed on an eFDP report.

Section 8. Source Reduction and Recycling Activities

The TRI-MEweb software offers a Section 8 Calculator. The Section 8 Calculator will assist users in calculating their Section 8 source reduction and recycling activity quantities. Please note that if you use range codes to report data in sections 5 and 6, TRI-MEweb will default to the mid-point of the range when performing section 8 calculations.

The entries in this section must be completed, even if your facility does not engage in source reduction or recycling activities.

Applies to: Paper forms and TRI-MEweb submissions

- Columns C and D, the future year projections for questions 8.1 through 8.7, must be completed. EPA expects a reasonable estimate for the future year projections. Zero can be used in columns C and D to indicate that the manufacture, process, or otherwise use of the chemical will be discontinued. In such cases, columns C and D for Section 8.1 through 8.7 must all contain zeroes.

Paper forms: Listed on an eFDP as a NOSE.

TRI-MEweb: TRI-MEweb submissions will not be allowed to be submitted to EPA with this error type.

Applies to: Paper forms only

- It is incorrect to use range codes to report quantities in Section 8. Range codes can be used only in Sections 5 and 6 of Form R.

- It is incorrect to use the same codes from Section 4 for reporting the maximum amount of the reported EPCRA Section 313 chemical on-site to report quantities in Section 8.

- Quantities reported in Section 8.1 through 8.7 are mutually exclusive and additive. This means that quantities of the reported EPCRA Section 313 chemical must not be double-counted in Section 8.1 through 8.7.

- Some double-counting errors have been due to confusion over the differences in how on-site treatment of an EPCRA Section 313 chemical is reported in Section 7A as compared to Section 8. In Section 7A, information on the treatment of *waste streams* containing the EPCRA Section 313 chemical is reported, along with the percent efficiency in terms of destruction **or** removal of the EPCRA Section 313 chemical from each waste stream. In Section 8, only the quantity of the *EPCRA Section 313 chemical* actually destroyed through the treatment processes reported in Section 7A is reported in Section 8.6 to avoid double-counting within Sections 8.1 through 8.7.

- Quantities reported in Section 8.1 through 8.7 must not be reported in Section 8.8 and vice versa.

- Any time a reported EPCRA Section 313 chemical is contained in a waste, and the waste is associated with routine production-related activities and is recycled, combusted for energy recovery, treated, disposed of, or otherwise released either on- or off-site, that quantity of the EPCRA Section 313 chemical must be included in the quantities reported in Sections 8.1 through 8.7

All calculation errors will be listed on an eFDP as a NOSE.

Reporting quantities in Section 8.1 Quantities of EPCRA Section 313 chemicals that are released (including disposed of) on-site and reported in Section 5 of Form R must be reported in either Section 8.1a or 8.1b.

$\S 8.1a = \S 5.4.1 + \S 5.5.1A + \S 5.5.1B - \S 8.8$ (on-site disposal to landfills or UIC Class I Wells)[1]

$\S 8.1b = \S 5.1 + \S 5.2 + \S 5.3 + \S 5.4.2 + \S 5.5.2 + \S 5.5.3A + \S 5.5.3B + \S 5.5.4 - \S 8.8$ (on-site disposal or other releases, other than disposal to landfills or UIC Class I Wells)[1]

Quantities of EPCRA Section 313 chemicals transferred off-site for the purposes of disposal reported in Section 6.2 using the following codes must appear in Section 8.1c:

- M64 Other Landfills
- M65 RCRA Subtitle C Landfills
- M81 Underground Injection to Class I Wells

$\S 8.1c = \S 6.1$ (portion of transfer that is untreated and ultimately disposed of in landfills or UIC Class I Wells) $+ \S 6.2$ (quantities associated with M codes M64, M65, and M81) $- \S 8.8$ (off-site disposal to landfills or UIC Class I Wells)[1]

Metals and metal category compounds transferred off-site to POTWs in Section 6.1 must appear in Section 8.1c or 8.1d. To report correctly in Sections 8.1a through d, a facility must include quantities that are disposed of or otherwise released to the environment either on-site or off-site, excluding disposal or other releases due to catastrophic events or non-production related activities.

Quantities of EPCRA Section 313 chemicals transferred off-site for the purposes of disposal reported in Section 6.2 using the following codes must appear in Section 8.1d:

- M10 Storage Only
- M41 Solidification/Stabilization - Metals and Metal Category Compounds Only
- M62 Wastewater Treatment (excluding POTW) - Metals and Metal Category Compounds Only
- M66 Subtitle C Surface Impoundment
- M67 Other Surface Impoundments
- M73 Land Treatment
- M79 Other Land Disposal
- M82 Underground Injection to Class II-V Wells
- M90 Other Off-Site Management
- M94 Transfer to Waste Broker - Disposal
- M99 Unknown.

§ 8.1d = § 6.1 (portion of transfer that is untreated and ultimately disposed of or otherwise released, other than disposal to landfills or UIC Class I Wells) + § 6.2 (quantities associated with M codes M10, M41, M62, M66, M67, M73, M79, M82, M90, M94, and M99) - § 8.8 (off-site disposal or other releases due to catastrophic events, other than disposal to landfills or UIC Class I Wells)[1]

All calculation errors will be listed on an eFDP as a NOSE.

Reporting quantities in Section 8.2 "Quantity used for energy recovery on-site." A quantity must be reported in Section 8.2 for the current (reporting) year when a method of on-site energy recovery is reported in Section 7B, and vice versa. An error facilities make when completing Form R is to report the methods of energy recovery used on-site in Section 7B but not report the total quantity associated with those methods. Another error is to report a quantity in this section if the combustion of the EPCRA Section 313 chemical took place in a system that did not recover energy (e.g., an incinerator). A quantity of the EPCRA Section 313 chemical combusted for energy recovery must not be reported if the EPCRA Section 313 chemical does not have a significant heating value. Examples of EPCRA Section 313 chemicals that do not have significant heating values include metals, metal portions of metal category compounds, and halons. Metals and metal portions of metal compounds will never be treated or combusted for energy recovery. Any quantities of the EPCRA Section 313 chemical associated with non-production related activities such as catastrophic releases and remedial actions, as well as other one-time events not associated with routine production practices that were combusted for energy recovery on-site must not be included in Section 8.8.

All calculation errors will be listed on an eFDP as a NOSE.

Reporting quantities in Section 8.3 "Quantity used for energy recovery off-site." As in Section 8.2, a quantity must not be reported in this section if the off-site combustion of the EPCRA Section 313 chemical took place in a system that did not recover energy (e.g., incinerator). A quantity of an EPCRA Section 313 chemical must not be reported as sent off-site for the purposes of energy recovery

if the EPCRA Section 313 chemical does not have a significant heating value. Examples of EPCRA Section 313 chemicals that do not have significant heating values include metals and metal portions of metal category compounds. Metals and metal portions of metal category compounds will never be combusted for energy recovery. Quantities must be reported in Section 8.3 that are reported in Section 6.2 as transferred off-site for the purposes of combustion for energy recovery using the following codes:

 - M56 Energy Recovery
 - M92 Transfer to Waste Broker - Energy Recovery

§ 8.3 = § 6.2 (energy recovery) - § 8.8 (off-site energy recovery)[2]

All calculation errors will be listed on an eFDP as a NOSE.

Reporting quantities in Section 8.4 "Quantity recycled on-site." A quantity must be reported in **Section 8.4** for the current reporting year when a method of on-site recycling is reported in Section 7C, and vice versa. An error a facility may make when completing Form R is to report the methods of recycling used on-site in Section 7C but not report the total quantity recovered using those methods.

In addition, only the amount of the chemical that was actually recovered is to be reported in Section 8.4. Any quantities of the EPCRA Section 313 chemical associated with non-production related activities such as catastrophic releases and remedial actions, as well as other one-time events not associated with routine production practices that were recycled on-site must not be included in Section 8.8.

All calculation errors will be listed on an eFDP as a NOSE.

Reporting quantities in Section 8.5. "Quantity recycled off-site." Quantities reported in Section 6.2 as transferred off-site for the purposes of recycling must be included in Section 8.5 using the following codes:

 - M20 Solvents/Organic Recovery
 - M24 Metals Recovery
 - M26 Other Reuse or Recovery
 - M28 Acid Regeneration

    – M93 Transfer to Waste Broker - Recycling.

§8.5 = §6.2 (recycling) - §8.8 (off-site recycling)[2]

All calculation errors will be listed on an eFDP as a NOSE.

Reporting quantities in Section 8.6 "Quantity treated on-site." Quantities may not always have to be reported in Section 8.6 when Section 7A is completed. This is because the information reported in Section 7A and Section 8 is different. Information on how waste streams containing the reported EPCRA Section 313 chemical are treated is reported in Section 7A, while the quantity of the EPCRA Section 313 chemical actually destroyed as a result of on-site treatment is reported in Section 8.6. If a quantity is reported in Section 8.6, Section 7A must be completed but the reverse may not be true. This may result in apparent discrepancies between Section 7A and Section 8. For example, a facility may treat wastewater containing an EPCRA Section 313 chemical by removing the EPCRA Section 313 chemical and then disposing of it on-site. The treatment of the wastewater would be reported in Section 7A, with an efficiency estimate based on the amount of the EPCRA Section 313 chemical removed from the wastewater. Although the chemical in the waste stream has been treated because the chemical has been removed, the EPCRA Section 313 chemical has not been treated because it has not been destroyed. The facility would report only the amount of the EPCRA Section 313 chemical actually destroyed during treatment in Section 8.6 and the amount ultimately disposed of in Section 8.1 to avoid double-counting the same quantity in Section 8. In cases where the EPCRA Section 313 chemical is not destroyed during a treatment process and subsequently enters another activity, such as disposal (e.g., metals removed from wastewater and subsequently disposed of on-site), the quantity of the EPCRA Section 313 chemical would be reported as disposed of in Section 8.1, not as treated in Section 8.6. Any quantities of the EPCRA Section 313 chemical associated with non-production related activities such as catastrophic releases and remedial actions, as well as other one-time events not associated with routine production practices that were treated for destruction on-site must not be included in Section 8.8. Metals generally will not be treated for destruction.

All calculation errors will be listed on an eFDP as a NOSE.

Reporting quantities in Section 8.7 "Quantity treated off-site." Quantities reported in Section 6.2 as transferred off-site for the purposes of

treatment must be included in Section 8.7 using the following codes:

    – M40 Solidification/Stabilization

    – M50 Incineration/Thermal Treatment

    – M54 Incineration/Insignificant Fuel Value

    – M61 Wastewater Treatment (excluding POTW)

    – M69 Other Waste Treatment

    – M95 Transfer to Waste Broker - Waste treatment.

Quantities of an EPCRA Section 313 chemical, except metals and metal category compounds, sent off-site to a POTW should also be reported in Section 8.7. If you know, however, that a chemical is not treated for destruction at the POTW you should report that quantity in Section 8.1 instead of 8.7.

To report correctly EPCRA Section 313 chemicals in Section 8.7, use the following equation.

§8.7 = §6.1 (portion of transfer that is ultimately treated) + §6.2 (treatment) - §8.8 (off-site treatment)[3]

All calculation errors will be listed on an eFDP as a NOSE.

Reporting quantities in Section 8.8 Quantity released to the environment as a result of remedial actions, catastrophic events or one-time events not associated with production processes. The quantities that are reported in Section 8.8 are associated with non-production related activities such as catastrophic releases and remedial actions, as well as one-time events not associated with routine production practices that were disposed of or released directly to the environment or transferred off-site for the purposes of recycling, energy recovery, treatment or disposal. Quantities reported in Section 8.8 must not be reported in Section 8.1 through 8.7.

Applies to: Paper forms and TRI-MEweb submissions

Reporting the production ratio in Section 8.9. A production ratio or activity index must be provided in Section 8.9. A zero is not acceptable and NA (Not Applicable) can be used only when the reported EPCRA Section 313 chemical was not manufactured, processed, or otherwise used in the year prior to the reporting year. TRI-MEweb in RY

---

[3]§8.8 includes quantities of toxic chemical disposed of or otherwise released on-site or managed as waste off-site due to remedial actions, catastrophic events, or one-time events not associated with the production processes.

2012 is providing an optional worksheet to help calculate the production ratio.

Calculating production ratio in Section 8.9. In calculating a production ratio for otherwise used chemicals, an activity index must be used rather than quantities purchased or released from year to year.

Reporting source reduction activities in Section 8.10. It is an error to report a source reduction activity in Section 8.10 and not report at least one method used to identify that activity and vice versa.

All calculation errors will be listed on an eFDP as a NOSE.

C.4 eFDP Messages Used to Report Notices of Significant Errors

Note: EPA is continually trying to improve the error checking system for TRI submissions. As a result, a small number of the error messages in this appendix may be changed by the time the Reporting Year 2012 submissions are checked. Most of these messages will remain the same. You can look for changes to these error messages on the TRI home page at http://www.epa.gov/tri

Applies to: Paper forms only

1. You have used an invalid Form R or Form A by using either a form not applicable for the reporting year, or a facsimile form that has not been approved by EPA. Resubmit your data on a current EPA approved Form R or A.

2. Pages were missing from the form received. Correct this by resubmitting a complete certified form for this chemical substance.

3. Multiple chemicals were reported in your Form R. You must submit a separate and complete Form R for each chemical cited.

4. You have provided a valid CAS number and a valid chemical name, but they do not match. Respond by providing a valid CAS number and matching chemical name.

5. You have left part or all of the chemical identification sections blank. Respond by providing a valid CAS number and matching chemical name or Mixture Component Identity.

6. You reported a CAS number and chemical name that are invalid. Respond by providing a valid CAS number and matching chemical name.

7. Your form indicated Trade Secret status with an indication that this form is a Sanitized version, but the report contains no Generic Chemical Name. You must provide a Generic Chemical Name for this sanitized form.

8. You have reported Dioxin and Dioxin-like Compounds on a Form A. Dioxin and Dioxin-like Compounds are not eligible for the alternate threshold. Thus, this chemical must be reported on a Form R. Please resubmit your data on a Form R.

9. In Part I, Section 1of the Form R or Form A Certification Statement You did not enter a reporting year. (Note: EPA has set the year to 2084 as a default.) You must enter a valid reporting year for your Form R or Form A Certification Statement. This entry cannot be left blank and NA may not be used. (NOSE)

10. In Part I, Section 1of the Form R or Form A Certification Statement you provided an invalid or future reporting year. You must enter a valid reporting year for your Form R or Form A Certification Statement. Valid years are 1987 through 2012. This entry cannot be left blank and NA may not be used. (NOSE)

11. You have reported a negative number(s) in Part II, Sections 5 and/or 6 and/or 8 of your Form R. Quantities reported in these sections must be 0 or greater. Please respond by providing correct release or other waste management data.

12. You did not complete Part II, Sections 5 and 6. Please provide the required information; otherwise indicate NA.

13. You did not complete Part II, Section 7. Please provide the required information; otherwise indicate NA.

14. You did not complete Part II, Section 8. Please provide the required information; otherwise indicate NA.

C.5 Messages Used to Report Notices of Technical Errors (NOTEs) and Non-technical Data Changes (NDCs)

Invalid codes throughout Form R

Applies to: Paper forms only

15. You submitted an invalid code. To correct this, consult the instructions for the proper table value and provide a valid code value. [Specific location on the form of the invalid code is given.] (NOTE)

16. PBT chemicals (e.g., Dioxin and Dioxin-like Compounds, Lead Compounds, Mercury Compounds and Polycyclic Aromatic Compounds (PACs)) are ineligible for range reporting for on-site releases and transfers off-site for further waste management. Please provide specific release,

transfer, and other waste management values.(NOTE)

17. For aluminum (fume or dust) or zinc (fume or dust), the Waste Management codes M56 and M92 are unacceptable. Please provide the proper Waste Management codes for these chemicals. (NOTE)

18. For asbestos (friable), the Waste Management codes M56 and M92 are unacceptable. Please provide the proper Waste Management codes for these chemicals. (NOTE)

General Errors for both the Form R and/or Form A

Applies to: Paper forms only

19. You reported a negative value for a release, transfer or other waste management quantity. Please provide a non-negative value for the specified part and section. (NOTE)

20. You have reported a value for a PBT chemical beyond seven digits to the right of the decimal. EPA's data management systems support data precision up to seven digits to the right of the decimal. EPA has truncated your numeric submission so the number of digits to the right of the decimal does not exceed seven. If this was incorrect, specify the correct value, not exceeding seven digits to the right of the decimal. (NDC)

Errors in Part I, Facility Identification Information

Applies to: Paper forms only

21. No selection was made in Part I, Section 2.1 and 2.2 (Trade Secret Information) and a generic chemical name was not provided in Part II, Section 1.3. Therefore, the No box was selected in Part I, Section 2.1. If this was incorrect, and you intended to make a trade secret claim of the identity of the toxic chemical, you must resubmit following the requirements of 40 CFR Part 350 to claim trade secret. (NDC)

22. You indicated trade secret in Part I, Section 2.1 (Trade Secret Information) but made no selection for Part I, Section 2.2 (sanitized/unsanitized) and did not provide a generic chemical name in Part II, Section 1.3. EPA changed your selection in Part I, Section 2.1 to indicate that a trade secret claim is not being made. If this was incorrect, and you intended to make a trade secret claim for the identity of the toxic chemical, you must resubmit following the requirements of 40 CFR Part 350 to claim trade secret. (NDC)

23. You made a selection of No in Part 1, Section 2.1 (Trade Secret Information) and selected unsanitized in Part 1, Section 2.2. In Part II, Section 1.3 a generic name was indicated. Part II, Section 1.3 should be completed only if trade secret is being claimed (Part 1, Section 2.1). EPA will move the chemical name information in Part II, Section 1.3 to Part II, Section 1.2. If this is incorrect and you wish to claim trade secret, you must resubmit following the requirements of 40 CFR Part 350. (NDC)

24. In Part I, Section 4.1, you entered NA or did not enter a county name, city name, state code, and/or zip code. These fields may not be left blank and NA is not an acceptable entry. You must provide a county name, city name, state code, and/or zip code where the facility is located. (NDC)

25. EPA has corrected the county name, city name, state code, and/ or zip code that you identified in Part I, Section 4.1. The county name, city name, state code, and/ or zip code that you identified was either misspelled, or incorrect, or did not match the previous year submissions. If you feel our correction was made in error, please resubmit forms with correct information. (NDC)

26. In Part I, Section 4.1, you have used an invalid TRIFID or you have self-assigned your own TRIFID or TRIFID that has been superseded. You may not generate your own TRIFID. The TRI Data Processing Center assigns this number to a facility. EPA has corrected this error and assigned you the correct TRIFID. Please note the corrected TRIFID and keep it for use in future submissions. (NDC)

27. No Public Contact name and/or telephone number was listed. Please provide the name and telephone number of your Public Contact. (NOTE)

28. No Technical Contact name and/or telephone number was listed. Please provide the name and telephone number of your Technical Contact. (NOTE)

29. The Federal Facility box was not checked on your form but we believe you are a Federal Facility. Unless you respond that you are not a Federal Facility, we will continue to treat you as a Federal Facility. (NOTE)

30. A valid NAICS code was not provided. Please provide at least one valid primary six-digit NAICS code. (NOTE)

31. You reported an invalid state code. If the address is in the US, please use a valid US Postal Service state code (see Table III of the Reporting Forms and Instructions). If the address is not in the US, please enter a valid code in the Country Field (see Table IV of the Reporting Forms and Instructions) (NOTE)

32. Either Box A (An Entire Facility) or Box B (Part of a Facility) should be checked in Part I, Section 4.2. One of the 2 boxes must be checked, but not both. (NOTE)

33. If applicable, check either Box C (Federal Facility) or Box D (GOCO) in Part I, Section 4.2, but do not check both boxes. (NOTE)

34. Dun and Bradstreet Numbers (Part I Section 4.6) are typically 9 characters in length. Please check the number(s) submitted. If they are incorrect, please make the appropriate changes. If you believe that they are correct, no further action is necessary. (NOTE)

35. If this is a North American phone number, please enter all 10 digits (i.e., include area code). If this is for another country, please begin the phone number with "011" as the prefix to your international telephone number. (NOTE)

36. In Part I, Section 3, you did not provide a printed or typed name and official title of owner/operator or senior management official. It cannot be N/A or left blank. Please provide a name for owner/operator or senior management official. (NOTE)

37. In Part I, Section 5.1 you did not enter the name of the parent company. This block cannot be left blank. You must enter the name for the parent company if it is a U.S. company. If it is a foreign company then you may check the [NA] box. (NOTE)

38. The parent company Dun and Bradstreet Number in Part I, Section 5.2 (typically a 9-digit number) cannot be left blank. However, if your parent company does not have a Dun and Bradstreet Number check the [NA] box next to Part I, Section 5.2. (NOTE)

Errors in Part II, Section 1. Toxic Chemical Identity

Applies to: Paper forms only

39. You have correctly identified the chemical but have used a synonym for the chemical name. EPA has changed the Chemical Name to use the preferred TRI nomenclature. Please specify the correct CAS Number and matching Chemical Name. (NDC)

40. The CAS number you reported was changed to match the chemical name reported, because the CAS number you provided was not a valid TRI Chemical. If this was incorrect, specify a valid CAS number and matching chemical name. (NDC)

41. The chemical name you reported was changed to match the CAS number reported, because the chemical name you provided was not a valid TRI Chemical. If this was incorrect, specify a valid CAS Number and matching Chemical Name. (NDC)

42. You reported a valid TRI CAS Number, a valid Chemical Name, and a generic Chemical Name. Therefore, the Generic Chemical Name was deleted. If this was incorrect, specify the Generic Chemical Name to be used. (NDC)

43. You reported a valid TRI CAS Number, a valid Chemical Name, and a Mixture Component Identity. Therefore, the Mixture Component Identity was deleted. If this was incorrect, specify the Mixture Component Identity to be used. (NDC)

44. EPA has changed the TRI chemical category code you reported in Part II, Section 1.1 from N151 to N150 (the code was incorrectly listed in some pages of the Reporting Forms and Instructions), the correct TRI chemical category code for Dioxin and Dioxin-like Compounds. If this is incorrect and you are not reporting Dioxin and Dioxin-like Compounds, please specify the correct CAS number or chemical category code and matching chemical name. (NDC)

45. You have reported for isopropyl alcohol (Only persons who manufacture by the strong acid process are subject) (CAS number 67-63-0). If you did not manufacture isopropyl alcohol by the strong acid process, you have submitted this form in error and should request that the form be withdrawn. (NOTE)

Errors in Part II, Section 3. Activities and Uses of Toxic Chemical at The Facility

Applies to: Paper forms only

46. You did not indicate in Part II, Section 3 which activity(ies) or use(s) of the EPCRA Section 313 chemical occur at your facility. Please indicate at least one of the activity(ies) and use(s) of the EPCRA Section 313 chemical occur at your facility. (NOTE)

Errors in Part II, Section 4. Maximum Amount of the Toxic Chemical Onsite at Any Time During the Calendar Year

Applies to: Paper forms only

47. You did not complete Part II, Section 4.1. Please provide a valid two digit code for the "maximum amount of chemical on-site at any time during the calendar year." (NOTE)

Errors in Part II, Section 5. Quantity of the Toxic Chemical Entering Each Environmental Medium Onsite

Applies to: Paper forms only

60. You did not complete Part II, Section 5.3. If you have discharged to water, please provide the Stream/Water Body name, the Release estimate or range code, Basis of Estimate and % from Stormwater; otherwise indicate "NA" (Not Applicable). (NOTE)

61. There are missing or incomplete data for Part II, Section 5.3. If you have discharged to water, please provide the Stream/Water Body name, the Release estimate or range code, Basis of Estimate and % from Stormwater; otherwise indicate "NA" (Not Applicable). (NOTE)

62. You did not complete Part II, Section 5. Please provide the Release estimate or range code and Basis of Estimate; otherwise indicate "NA" (Not Applicable). (NOTE)

63. There are missing or incomplete data for Part II, Section 5. Please provide the Release estimate or range code and Basis of Estimate; otherwise indicate "NA" (Not Applicable). (NOTE)

Errors in Part II, Section 6. Transfers of the Toxic Chemical in Wastes To Off-Site Locations

Applies to: Paper forms only

64. You did not complete Part II, Section 6.1, "discharges to POTW." If you did not discharge wastewater containing the Section 313 chemical to a POTW(s), enter "NA" (Not Applicable), otherwise please provide the Transfer amount or range code, Basis of Estimate, POTW Name and Location. (NOTE)

65. You reported a POTW(s) name and location but did not provide a Transfer amount. Please provide a Total Transfer amount or range code and Basis of Estimate; otherwise, if there was no transfer to a POTW of wastewater that contains or contained the Section 313 chemical, delete the POTW location and indicate "NA" (Not Applicable) for the POTW transfer amount. (NOTE)

66. You reported a Total Transfer amount or range code and Basis of Estimate in Part II Section 6.1 but did not indicate a POTW name and location in Section 6.1.B. Please provide the POTW Name and Location. (NOTE)

67. You provided an incomplete POTW name and address. Please provide the name and complete address for the POTW. (NOTE)

68. There are missing or incomplete data for Part II, Section 6.1. Please provide the transfer amount or range code and Basis of Estimate for Discharges to POTWs. (NOTE)

69. You did not complete Part II, Section 6.2, "Transfers to Other Off-site Locations." If you did not transfer the waste containing the Section 313 chemical to other off-site locations, enter "NA" (Not Applicable), otherwise please provide Offsite EPA ID, Name, Location, Transfer amount or range code, Basis of Estimate, and type of Waste Management code. (NOTE)

70. You reported an Off-site Transfer amount or range code and Basis of Estimate in Part II Section 6.2 but did not indicate an Off-site name and location in Section 6.2. Please provide the Off-site Name and Location. (NOTE)

71. You reported an Off-site name and location but did not provide a Transfer amount. Please provide a Total Transfer amount or range code, Basis of Estimate and type of Waste Management code; otherwise, if there was no transfer to this Off-site location, delete the Off-site name and location and indicate "NA" (Not Applicable) in the Off-site EPA Identification Number (RCRA ID No.) field. (NOTE)

72. You provided both county and country data. If this is an extra-national transfer, indicate the off-site name, address, and Country Code; if a domestic Offsite, provide the Off-site Name and correct address. (NOTE)

73. You reported an Off-site name and location, but there are missing or incomplete data for the off-site transfer amount, basis of estimate and type of waste management code. Please provide the Off-site Transfer amount or range code, Basis of Estimate, and type of Waste Management code. (NOTE)

74. You provided incomplete off-site name and address data. For a transfer to a domestic off-site location, you must provide a street address, city, state, county and zip code. For a transfer to a foreign off-site location, you must provide a street address, city and a two character country code. (NOTE)

75. You reported an invalid Type of Waste Management code. For metals/metal compounds use only disposal and certain recycling activities codes. Consult the Reporting Instructions for metal and metal compounds and correct with a valid Waste Management (i.e., "M") code. (NOTE)

76. You reported an invalid Type of Waste Management code. For Barium Compounds use only disposal and certain recycling activities codes, M61-Wastewater Treatment (Excluding POTW) or M69-Other Waste Treatment. Consult the Reporting Instructions for metal and metal compounds and correct with a valid Waste Management (i.e., "M") code. (NOTE)

77. For non-metals codes M41 and M62 are unacceptable. Provide the appropriate Disposal or Other Waste Management code for this non-metal substance. (NOTE)

78. In Part II, Section 6.2 column C you reported M codes (M56 and/or M92) for energy recovery, however you left Section 8.3 column B blank. Please provide the quantity used for energy recovery offsite in pounds/year in Section 8.3 column B. (NOTE)

79. In Part II, Section 6.2 column C you reported M Codes (M20, M24, M26, M28, M93) for recycling, however you left Section 8.5 column B blank. Please provide the quantity recycled offsite in pounds/year in Section 8.5 column B. (NOTE)

80. In Part II, Section 6.2 column C you reported M Codes (M40, M50, M54, M61, M69, M95) for treatment, however you left Section 8.7 column B blank. Please provide the quantity treated offsite in pounds/year in Section 8.7 column B. (NOTE)

Errors in Part II, Section 7. On-Site Waste Treatment Methods and Efficiency

Applies to: Paper forms only

81. There are no data contained in all of Part II, Section 7A. If you do not treat wastes containing the EPCRA Section 313 chemical at your facility, indicate "NA;" otherwise please provide the general waste stream code, waste treatment methods, range of influent concentration, waste treatment efficiency estimate and whether this is based on operating data for all on-site waste treatments for this chemical. (NOTE)

82. There are missing data in Part II, Section 7A. Please provide the general waste stream code, waste treatment methods, range of influent concentration, waste treatment efficiency estimate and whether this is based on operating data. (NOTE)

83. There are no data in Part II, Section 7B. If no on-site energy recovery processes are used for this Section 313 chemical at your facility, indicate "NA;" otherwise please provide at least one three-character on-site energy recovery process code. (NOTE)

84. There are no data in Part II, Section 7C. If no on-site recycling processes are used for this Section 313 chemical at your facility, indicate "NA;" otherwise please provide at least one three-character on-site recycling process code. (NOTE)

Errors in Part II, Section 8. Source Reduction and Recycling Activities

Applies to: Paper forms only

85. There are missing data for Part II, Section 8.1-8.7. Please provide an estimate or "NA" (Not Applicable) in each box for section 8.1 through 8.7, columns A, B, C, and D. You may only use "NA" (Not Applicable) when there is no possibility a release or transfer occurred. You may enter zero if the release or transfer was equal to or less than half a pound. (NOTE)

86. There are missing data in Part II, Section 8.8. Please provide an estimate or "NA" (Not Applicable). You may only use "NA" (Not Applicable) when there is no possibility a release or transfer occurred. You may enter zero if the release or transfer was equal to or less than half a pound. (NOTE)

87. There are no data in Part II, Section 8.9. Please provide a production ratio, an activity index, or "NA" (Not Applicable) if the chemical manufacture or use began during the current reporting year. (NOTE)

88. There are no data in Part II, Section 8.10. If your facility did not engage in any source reduction activity for the reported chemical, enter "NA" (Not Applicable) and answer 8.11. Otherwise please provide Source Reduction Activities and Methods code(s). (NOTE)

89. There are missing data in Part II, Section 8.10. Please provide Source Reduction Activities and Methods code(s). (NOTE)

90. You have reported a listed metal or metal compound category in section 8.2, 8.3, 8.6 or 8.7. However, these chemicals cannot be treated for destruction. Metal or metal compound category can only be reported as disposed of or recycled. Please report appropriately in Section 8.1, 8.4, or 8.5. (NOTE)

91. You reported a negative value for a release, transfer or other waste management quantity. Please provide a non-negative value for the specified part and section. (NOTE)

Errors relating to the reconciliation of data in Part II, Section 8 and Part II, Sections 5, 6, and 7

Applies to: Paper forms only

92. You did not complete Sections 8.1 through 8.7 column B or 8.8. If you report releases in Part II, Section 5 and/or an off-site transfer in Section 6.2 and/or quantities transferred off-site to POTWs in Section 6.1, you must report an estimate in Part II, Sections 8.1 through 8.7 column B and/or Section 8.8. (NOTE)

93. You did not complete Sections 5, 6, or 7. If you enter an estimate in Part II, Sections 8.1 through 8.7, column B and/or Section 8.8, you must also report releases in Part II, Section 5 and/or off-site transfers in Section 6.2 and/or quantities transferred off-site to POTWs in Section 6.1 and/or waste treatment, energy recovery, or recycling codes in Section 7. Please provide data for Sections 5, 6, and/or 7. (NOTE)

94. You reported an estimate in Part II, Section 8.2, column B, "Quantity Used for Energy Recovery On-site," but did not provide an on-site energy recovery code in Part II, Section 7B. Please provide an on-site energy recovery code for Part II, Section 7B. (NOTE)

95. You reported an "On-site Energy Recovery Process" code in Part II, Section 7B, but you did not provide an estimate of the quantity used for energy recovery in Part II, Section 8.2, column B. Please provide an estimate of the quantity used for energy recovery for Part II, Section 8.2, column B. (NOTE)

96. You reported an estimate in Part II, Section 8.4, column B "Quantity Recycled On-site" but did not provide an on-site recycling code in Part II, Section 7C. Please provide an on-site recycling code for Part II, Section 7C. (NOTE)

97. You reported one or more on-site recycling process codes in Part II, Section 7C but did not provide an estimate in Part II, Section 8.4, column B, "Quantity Recycled On-site." Please provide an estimate of the quantity recycled for Section 8.4 column B. (NOTE)

98. You reported a value in Part II, Section 8.3 column B, however you did not provide a corresponding quantity with an appropriate M Code (M56 and/orM92) for energy recovery in Section 6.2 column C. Please provide the appropriate quantity and M Codes for energy recovery in Section 6.2 column C. (NOTE)

99. You reported a value in Part II, Section 8.5 column B, however you did not provide a corresponding quantity with an appropriate M Code (M20, M24, M26, M28, M93) for recycling in Section 6.2 column C. Please provide the appropriate quantity and M Codes for recycling in Section 6.2 column C. (NOTE)

100. You reported a value in Part II, Section 8.7 column B, however you did not report a quantity in Section 6.1 or a quantity with an appropriate M Code (M40, M50, M54, M61, M69, M95) for treatment in Section 6.2 column C. Please provide a quantity in Section 6.1 or the appropriate quantity and M Codes for treatment in Section 6.2 column C. (NOTE)

101. You have reported a listed metal or metal compound category in Part II, Section 6.1, however you have not provided a quantity released in section 8.1 column B. Note that in Section 8a, metal or metal compound category can only be reported as disposed of or recycled and not reported as treated for energy recovery or treated for destruction. Please provide quantity released in pounds/year in Section 8.1 column B. (NOTE)

102. You have reported a listed metal or metal compound category in Part II, Section 6.1, however you have not provided quantity released in 8.1d Column B. Note that in Section 8a, metal or metal compound category can only be reported as disposed of or recycled and not reported as treated for energy recovery or treated for destruction. Please provide quantity released in pounds/year Section 8.1B. (NOTE)

Appendix D. Supplier Notification Requirements

EPA requires some suppliers of mixtures or other trade name products containing one or more of the EPCRA section 313 chemicals to notify their customers. This requirement has been in effect since January 1, 1989.

This appendix explains which suppliers must notify their customers, who must be notified, what form the notice must take, and when it must be sent.

D.1 Who Must Supply Notification

You are covered by the section 313 supplier notification requirements if you own or operate a facility which meets all of the following criteria:

1. Your facility is in a North American Industry Classification System (NAICS) code that corresponds to Standard Industrial Classification [SIC] codes 20–39;

2. You manufacture (including import) or process an EPCRA section 313 chemical; and

3. You sell or otherwise distribute a mixture or other trade name product containing the EPCRA section 313 chemical to either:

    – A facility in a covered NAICS code (see Table I).

    – A person that then may sell the same mixture or other trade name product to a firm in a covered NAICS code (see Table I).

Note that you may be covered by the supplier notification rules even if you are not covered by the section 313 release reporting requirements. For example, even if you have fewer than 10 full-time employees or do not manufacture or process any of the EPCRA section 313 chemicals in sufficient quantities to trigger the release and other waste management reporting requirements, you may still be required to notify certain customers.

D.2 Who Must Be Notified

Industries whose primary NAICS code does not correspond to SIC codes 20 through 39 are not required to initiate the distribution of notifications for EPCRA section 313 chemicals in mixtures or other trade name products that they send to their customers.

However, if these facilities receive notifications from their suppliers about EPCRA section 313 chemicals in mixtures or other trade name products, they should forward the notifications with the EPCRA section 313 chemicals they send to other covered users.

An example would be if you sold a lacquer containing toluene to distributors who then may sell the product to other manufacturers. The distributors are not in a covered NAICS code, but because they sell the product to companies in covered NAICS codes, they must be notified so that they may pass the notice along to their customers, as required.

The language of the supplier notification requirements covers mixtures or other trade name products that are sold or otherwise distributed. The "otherwise distributes" language includes intra-company transfers and, therefore, the supplier notification requirements at 40 CFR Section 372.45 apply.

D.3 Supplier Notification Content

The supplier notification must include the following information:

1. A statement that the mixture or other trade name product contains an EPCRA section 313 chemical or chemicals subject to the reporting requirements of EPCRA section 313 (40 CFR 372);

2. The name of each EPCRA section 313 chemical and the associated Chemical Abstracts Service (CAS) registry number of each chemical if applicable. (CAS numbers are not used for chemical categories, since they can represent several individual EPCRA section 313 chemicals.); and

3. The percentage, by weight, of each EPCRA section 313 chemical (or all EPCRA section 313 chemicals within a listed category) contained in the mixture or other trade name product.

For example, if a mixture contains a chemical (i.e., 12 percent zinc oxide) that is a member of a reportable EPCRA section 313 chemical category (i.e., zinc compounds), the notification must indicate that the mixture contains a zinc compound at 12 percent by weight. Supplying only the weight percent of the parent metal (zinc) does not fulfill the requirement. The customer must be told the weight percent of the entire compound within an EPCRA section 313 chemical category present in the mixture.

D.4 How the Notification Must Be Made

The required notification must be provided at least annually in writing. Acceptable forms of notice include letters, product labeling, and product literature distributed to customers. If you are required to prepare and distribute a Material Safety Data Sheet (MSDS) for the mixture under the Occupational Safety and Health Act (OSHA) Hazard Communication Standard, your section 313 notification must be attached to the MSDS or the MSDS must be modified to include the required information. (A sample letter and recommended text for inclusion in an MSDS appear at the end of this appendix.)

You must make it clear to your customers that any copies or redistribution of the MSDS or other form of notification must include the section 313 notice. In other words, your customers should understand their requirement to include the section 313 notification if they give your MSDS to their customers.

D.5 When Notification Must Be Provided

You must notify each customer receiving a mixture or other trade name product containing an EPCRA section 313 chemical with the first shipment of each calendar year. You may send the notice with subsequent shipments as well, but it is required that you send it with the first shipment each year. Once customers have been provided with an MSDS containing the section 313 information, you may refer to the MSDS by a written letter in subsequent years (as long as the MSDS is current).

If EPA adds EPCRA section 313 chemicals to the section 313 list, and your products contain the newly added EPCRA section 313 chemicals, notify your customers with the first shipment made during the next calendar year following EPA's final decision to add the chemical to the list. For example, if EPA adds chemical ABC to the list in September 1998, supplier notification for chemical ABC would have begun with the first shipment in 1999.

You must send a new or revised notice to your customers if you:

1. Change a mixture or other trade name product by adding, removing, or changing the percentage by weight of an EPCRA section 313 chemical; or

2. Discover that your previous notification did not properly identify the EPCRA section 313 chemicals in the mixture or correctly indicate the percentage by weight.

In these cases, you must:

1. Supply a new or revised notification within 30 days of a change in the product or the discovery of misidentified EPCRA section 313 chemical(s) in the mixture or incorrect percentages by weight; and

2. Identify in the notification the prior shipments of the mixture or product in that calendar year to which the new notification applies (e.g., if the revised notification is made on August 12, indicate which shipments were affected during the period January 1–August 12).

D.6 When Notifications Are Not Required

Supplier notification is not required for a "pure" EPCRA section 313 chemical unless a trade name is used. The identity of the EPCRA section 313 chemical will be known based on label information.

You are not required to make a "negative declaration." That is, you are not required to indicate that a product contains no EPCRA section 313 chemicals.

If your mixture or other trade name product contains one of the EPCRA section 313 chemicals, you are not required to notify your customers if:

1. Your mixture or other trade name product contains the EPCRA section 313 chemical in percentages by weight of less than the following levels (These are known as *de minimis* levels)

    –   0.1 percent if the EPCRA section 313 chemical is defined as an "OSHA carcinogen;"

    –   1 percent for other EPCRA section 313 chemicals.

De minimis levels for each EPCRA section 313 chemical and chemical category are listed in Table II. PBT chemicals (except lead when contained in stainless steel, brass or bronze alloys) are not eligible for the *de minimis exemption*. Therefore, *de minimis* levels are not provided for these chemicals in Table II. However, for purposes of supplier notification requirements only, such notification is not required when the following PBT chemicals are contained in mixtures below their respective *de minimis* levels:

Chemical or chemical category name	CAS number or chemical category code	Supplier notification limit (%)
Aldrin	309-00-2	1.0
Benzo[g,h,i]perylene	191-24-2	1.0
Chlordane	57-74-9	0.1
Dioxin and dioxin-like compounds (manufacturing; and the processing or otherwise use of dioxin and dioxin-like compounds if the dioxin and dioxin-like compounds are present as contaminants in a chemical and if they were created during the manufacturing of that chemical	N150	1.0*
Heptachlor	76-44-8	0.1
Hexachlorobenzene	118-74-1	0.1
Isodrin	465-73-6	1.0
Lead	7439-92-1	0.1
Lead compounds	N420	0.1**
Mercury	7439-97-6	1.0
Mercury compounds	N458	1.0
Methoxychlor	72-43-5	1.0
Octachlorostyrene	29082-74-4	1.0
Pendimethalin	40087-42-1	1.0
Pentachlorobenzene	608-93-5	1.0
Polychlorinated biphenyls (PCBs)	1336-36-3	0.1
Polycyclic aromatic compounds category	N590	0.1***
Tetrabromobisphenol A	79-94-7	1.0
Toxaphene	8001-35-2	0.1

Chemical or chemical category name	CAS number or chemical category code	Supplier notification limit (%)
Trifluralin	1582-09-8	1.0

*The *de minimis* level is 1.0 for all members except for 2,3,7,8-Tetrachlorodibenzo-*p*-dioxin which has a 0.1% *de minimis* level.

**The *de minimis* level is 0.1 for inorganic lead compounds and 1.0 for organic lead compounds

***The *de minimis* level is 0.1 except for benzo(a)phenanthrene, dibenzo(a,e)fluoranthene, benzo(j,k)fluorene, and 3-methylcholanthrene which are subject to the 1.0% *de minimis* level.

2.   Your mixture or other trade name product is one of the following:

–   An article that does not release an EPCRA section 313 chemical under normal conditions of processing or otherwise use.

–   Foods, drugs, cosmetics, alcoholic beverages, tobacco, or tobacco products packaged for distribution to the general public.

–   Any consumer product, as the term is defined in the Consumer Product Safety Act, packaged for distribution to the general public. For example, if you mix or package one-gallon cans of paint designed for use by the general public, notification is not required.

3.   A waste sent off site for further waste management. The supplier notification requirements apply only to mixtures and trade name products. They do not apply to wastes.

4.   You are initiating distribution of a mixture or other trade name product containing one or more EPCRA section 313 chemicals and your facility is in any of the covered SIC codes added during the 1997 industry expansion rulemaking, including facilities whose SIC code is within SIC major group codes 10 (except 1011, 1081, and 1094), 12 (except 1241); industry codes 4911 (limited to facilities that combust coal and/or oil for the purpose of generating power for distribution in commerce), 4931 (limited to facilities that combust coal and/or oil for the purpose of generating power for distribution in commerce), or 4939 (limited to facilities that combust coal and/or oil for the purpose of generating power for distribution in commerce); or 4953 (limited to facilities regulated under the

Resource Conservation and Recovery Act, subtitle C, 42 U.S.C. Section 6921 et seq.) or 5169, or 5171, or 7389 (limited to facilities primarily engaged in solvents recovery services on a contract or fee basis).

D.7 Trade Secrets

Chemical suppliers may consider the chemical name or the specific concentration of an EPCRA section 313 chemical in a mixture or other trade name product to be a trade secret. If they consider:

1. The specific identity of an EPCRA section 313 chemical to be a trade secret, the notice must contain a generic chemical name that is descriptive of the structure of that EPCRA Section 313 chemical (for example, decabromodiphenyl oxide could be described as a halogenated aromatic);

2. The specific percentage by weight of an EPCRA section 313 chemical in the mixture or other trade name product to be a trade secret, the notice must contain a statement that the EPCRA section 313 chemical is present at a concentration that does not exceed a specified upper bound. For example, if a mixture contains 12 percent toluene and you consider the percentage a trade secret, the notification may state that the mixture contains toluene at no more than 15 percent by weight. The upper

bound value chosen must be no larger than necessary to adequately protect the trade secret.

If you claim this information to be trade secret, you must have documentation that provides the basis for your claim.

D.8 Recordkeeping Requirements

You are required to keep records of the following for three years:

1. Notifications sent to recipients of your mixture or other trade name product;

2. All supporting materials used to develop the notice;

3. If claiming a specific EPCRA section 313 chemical identity a trade secret, you should record why the EPCRA section 313 chemical identity is considered a trade secret and the appropriateness of the generic chemical name provided in the notification; and

4. If claiming a specific concentration a trade secret, you should record explanations of why a specific concentration is considered a trade secret and the basis for the upper bound concentration limit.

Information retained under 40 CFR 372 must be readily available for inspection by EPA.

D.9 Sample Notification Letter

January 2, 2009

Mr. Edward Burke
Furniture Company of North Carolina
1000 Main Street
Anytown, North Carolina 99999

Dear Mr. Burke:

This letter is to inform you that a product that we sell to you, Furniture Lacquer KXZ-1390, contains one or more chemicals subject to section 313 of Emergency Planning and Community Right-to-Know Act (EPCRA). We are required to notify you of the presence of these chemicals in the product under EPCRA section 313. This law requires certain industrial facilities to report on annual emissions and other waste management of specified EPCRA section 313 chemicals and chemical categories. Our product contains:

   Toluene, Chemical Abstract Service (CAS) number 108-88-3, 20 percent, and

   Zinc compounds, 15 percent.

If you are unsure whether you are subject to the reporting requirements of EPCRA section 313, or need more information, call the EPA/TRI Information Center. For contact information, please see the TRI Home Page at http://www.epa.gov/tri. Your other suppliers should also be notifying you about EPCRA section 313 chemicals in the mixtures and other trade name products they sell to you.

Finally, please note that if you repackage or otherwise redistribute this product to industrial customers, a notice similar to this one should be sent to those customers.

                              Sincerely,

                              Emma Sinclair
                              Sales Manager
                              Furniture Products

D.10 Sample Notification on an MSDS Furniture Products

Section 313 Supplier Notification

This product contains the following EPCRA section 313 chemicals subject to the reporting requirements of section 313 of the Emergency Planning and Community Right-To-Know Act of 1986 (40 CFR 372):

CAS Number	Chemical Name	Percent by Weight
108-88-3	Toluene	20%
NA	Zinc Compounds	15%

This information must be included in all MSDSs that are copied and distributed for this material.

Material Safety Data Sheet

Appendix E.   TRI State and Tribal Contacts

EPCRA Section 313 requires facilities to submit reports to both EPA and their state or tribe (if located in Indian country as defined by 18 USC §1151). For a current list of state and tribal designated Section 313 contacts, see the TRI web site at:

State TRI Contact Information and Web Sites: http://www.epa.gov/tri/stateprograms/state_programs.htm

Tribal TRI Contact Information and Web Sites: http://www.epa.gov/tri/contacts/contacts_tribal.html

Appendix F. TRI Regional Contacts

Region 1 (CT, ME, MA, NH, RI, and VT)

Dwight Peavey
Assistance and Pollution Prevention Office
USEPA Region 1 (SPT)
1 Congress Street, Suite 1100
Boston, MA 02114-2023
(617) 918-1829; fax: (617) 918-1810
peavey.dwight@epa.gov

Region 2 (NJ, NY, PR, and VI)

Nora Lopez
Pesticides and Toxic Substances Branch
USEPA Region 2 (MS-105)
2890 Woodbridge Avenue, Building 10
Edison, NJ 08837-3679
(732) 906-6890; fax: (732) 321-6788
lopez.nora@epa.gov

Region 3 (DE, DC, MD, PA, VA, and WV)

William Reilly
Toxics Programs and Enforcement Branch
USEPA Region 3 (3WC33)
1650 Arch Street
Philadelphia, PA 19103-2029
(215) 814-2072; fax: (215) 814-3114
reilly.william@epa.gov

Region 4 (AL, FL, GA, KY, MS, NC, SC, TN)

Ezequiel Velez
EPCRA Enforcement Section
USEPA Region 4
Atlanta Federal Center
61 Forsyth Street, S.W.
Atlanta, GA 30303-8960
(404) 562-9191; fax: (404) 562-9163
velez.ezequiel@epa.gov

Region 5 (IL, IN, MI, MN, OH, and WI)

Thelma Codina
Pesticides and Toxics Branch
USEPA Region 5 (DT-8J)
77 West Jackson Boulevard
Chicago, IL 60604
(312) 886-6219; fax: (312) 353-4788
codina.thelma@epa.gov

Region 6 (AR, LA, NM, OK, and TX)

Morton Wakeland
Toxics Section, Multimedia Planning and
  Permitting Division
USEPA Region 6 (6PD-T)
1445 Ross Avenue, Suite 1200
Dallas, TX 75202-2733
(214) 665-8116; fax: (214) 665-6762
wakeland.morton@epa.gov

Region 7 (IA, KS, MO, and NE)

Stephen Wurtz
Air, RCRA and Toxics Division
USEPA Region 7 (ARTD/CRIB)
901 North 5th Street
Kansas City, KS 66101
(913) 551-7315; fax: (913) 551-7065
wurtz.stephen@epa.gov

Region 8 (CO, MT, ND, SD, UT, and WY)

Barbara Conklin
Office of Pollution Prevention, Pesticides and
  Toxics
USEPA Region 8 (8P-P3T)
1595 Winkoop Street
Denver, CO 80202
(303) 312-6619; fax: (303) 312-6044
conklin.barbara@epa.gov

Region 9 (AS, AZ, CA, GU, HI, MH, MP, and NV)

Lily Lee
Toxics Office, Communities and Ecosystems
  Division
USEPA Region 9 (CED-4)
75 Hawthorne Street
San Francisco, CA 94105-3901
(415) 947-4187; fax: (415) 947-3583
lee.lily@epa.gov

Region 10 (AK, ID, OR, and WA)

Gabriela Carvalho
USEPA Region 10 (AWT-128)
1200 Sixth Avenue
Suite 900
Seattle, WA 98101-3140
(206) 553-4016; fax: (206) 553-7176
carvalho.gabriela@epa.gov

Appendix G. Other Relevant Section 313 Materials

G.1 TRI National Analysis

2012 Toxics Release Inventory National Analysis

EPA summarizes the latest TRI data in a report called the TRI National Analysis. The National Analysis is an annual report that includes information about toxic chemical releases to the environment, how toxic chemicals are managed at TRI facilities (i.e. recycled, treated and burned for energy), and how facilities are working to reduce toxic chemicals they generate and release. The TRI National Analysis Overview document includes national trends and figures, while other websites linked from the National Analysis homepage include more localized analyses of states, certain urban areas and watersheds. The National Analysis homepage can be accessed at www.epa.gov/tri/NationalAnalysis.

To conduct your own analysis, TRI data collected from 1987 through 2011 can be accessed using the TRI Explorer online tool:

http://www.epa.gov/triexplorer

as well as several other public access tools available on the TRI website at http://www.epa.gov/tri/tridata/index.htm.

G.2. Access to TRI Information On-line

The **TRI Home Page** http://www.epa.gov/tri offers information useful to both novice and experienced users of the Toxics Release Inventory. It provides a description of what the TRI database is and how it can be used; access to TRI data; TRI regulations; and guidance documents for complying with TRI regulations and using TRI data. You can find out about TRI products, view or download the 2012 TRI reports, and identify who to contact for more information in EPA regions and state programs across the country. From the TRI home page, you can link to other EPA and non-EPA sites that also allow you to search the TRI database and other databases online.

TRI Explorer http://www.epa.gov/triexplorer is an on-line tool that EPA has created to obtain TRI data. It allows the user to search the TRI database using six criteria: facility, chemical, year or industry type (NAICS code), federal facility and geographic area (at the county, state or national level). The tool will generate three types of reports: (1) Release Reports (including on- and off-site releases (i.e., off-site releases include transfers off-site to disposal and metals and metal compounds transferred to POTWs)); (2) Waste Transfer Reports (including amounts transferred off-site for further waste management but not including transfers off-site to disposal); and (3) Waste Quantity Reports (including amounts recycled, burned for energy recovery, quantities treated, and quantities released).

TOXNET http://toxnet.nlm.nih.gov the National Library of Medicine's (NLM) Toxicology Data Network, provides free access to several databases, including the TRI database, that provides a variety of information on toxic chemicals. As with EPA's TRI Explorer tool, users of TOXNET can search by chemical or other name, chemical name fragment, or Chemical Abstracts Service Registry Number. Also searchable are facility or parent company name, state, city, county, or zip code. Search results can be limited to releases greater than a specified number of pounds, and individual releases can be summed together to display a total amount. Toxicity and environmental fate data for thousands of chemicals are also available from TOXNET.

G.3 Other TRI Information

EPA's Integrated Risk Information System (IRIS) http://www.epa.gov/iris is an electronic database containing information on human health effects that may result from exposure to various chemicals, including TRI chemicals, in the environment. IRIS was initially developed for EPA staff in response to a growing demand for consistent information of chemical substances for use in risk assessments, decision-making and regulatory activities. The information in IRIS is intended for those without extensive training in toxicology, but with some knowledge of health sciences.

Consolidated List of Chemicals Subject to the Emergency Planning and Community Right-to-Know Act and Section 112(r) of the Clean Air Act (List of Lists), (October 2001):

http://www.epa.gov/ceppo/pubs/title3.pdf

A paper copy is available from the National Technical Information Service, 5285 Port Royal Road, Springfield, VA 22161, 703 605-6000, Document Number: PB2003-105834, $38.00 plus

$5.00 shipping and handling.

Chemicals in Your Community, A Citizen's Guide to the Emergency Planning and Community Right-to-Know Act, December 1999 (EPA 550-99-001).

This booklet is intended to provide a general overview of the EPCRA requirements and benefits for all audiences. Part I of the booklet describes the provisions of EPCRA and Part II describes more fully the authorities and responsibilities of groups of people affected by the law and is available through written request at no charge from the EPA/TRI Information Center. For contact information, please see the TRI Home Page at http://www.epa.gov/tri.

Chemicals in the Environment

Issue number 6 of Chemicals in the Environment (CIE), published in the Fall of 1997, is devoted entirely to TRI. This 22 page publication contains 19 articles ranging from the history of TRI to the future of new TRI products. Articles include perspectives from the community, state, Federal, and International level. The publication also provides valuable information on training and contacts within the EPA. CIE is available free from EPA by asking for publication EPA 749-R-97-001b. To request copies, contact:

> U.S. Environmental Protection Agency
> Ariel Rios Building
> 1200 Pennsylvania Ave., N.W.
> Attn: TRI Documents
> MC: 2844
> Washington, DC 20460
>
> 202 564-9554
> Email: TRIDOCS@epa.gov

The Pollution Prevention Information Clearinghouse (PPIC)

> http://www.epa.gov/oppt/ppic/index.html

PPIC was established as part of EPA's response to the Pollution Prevention Act of 1990, which directed the Agency to compile information, including a database, on management, technical, and operational approaches to source reduction. PPIC provides information to the public and industries involved in conservation of natural resources and in reduction or elimination of pollutants in facilities, workplaces, and communities.

To request EPA information on pollution prevention or obtain fact sheets on pollution prevention from

various state programs call the PPIC reference and referral service at 202 566-0799, or fax a request to 202 566-0794, or write to:

> U.S. EPA
> Pollution Prevention Information Clearinghouse (PPIC)
> EPA West
> 1200 Pennsylvania Ave. NW
> Room 3379 (Mail Code 7407-T)
> Washington, DC 20460-0001

Email: ppic@epa.gov

Appendix H.   Guidance Documents

H.1 General Guidance

Many of the TRI guidance documents are available via the Internet http://www.epa.gov/tri.

- **40 CFR 372, Toxic Chemical Release Reporting; Community Right-to-Know; Final Rule**
  A reprint of the final EPCRA section 313 rule as it appeared in the *Federal Register* (FR) February 16, 1988 (53 FR 4500) (OTSFR 021688).

- **Common Synonyms for Chemicals Listed Under Section 313 of the Emergency Planning and Community Right-to-Know Act**
  March 1995 (EPA 745R-95-008)

  This glossary contains chemical names and their synonyms for substances covered by the reporting requirements of EPCRA section 313. The glossary was developed to aid in determining whether a facility manufactures, processes, or otherwise uses a chemical subject to EPCRA section 313 reporting.

- **EPCRA Section 313 Questions and Answers - Revised 1998 Version**
  December 1998 (EPA 745-B-98-004)

  The revised 1998 *EPCRA Section 313 Questions and Answers* document assists regulated facilities in complying with the reporting requirements of EPCRA section 313. This updated document presents interpretive guidance in the form of answers to many commonly asked questions on compliance with EPCRA section 313. In addition, this document includes comprehensive written directives to assist covered facilities in understanding some of the more complicated regulatory issues. This updated guidance document is intended to supplement the instructions for completing the Form R and the Alternate Threshold Certification Statement (Form A).

- **EPCRA Section 313 Questions and Answers - Addendum to the Revised 1998 Version**
  December 2004 (EPA-260-B-04-002)

  As a result of Executive Order 13148, regulatory actions, and legal decisions over the past five years, some of the Qs & As contained in the 1998 Q &A Document were updated. The 1998 Q & A Document remains valid guidance in all other respects.

- **EPCRA Section 313 Questions and Answers Addendum for Federal Facilities**
  May 2000 (EPA 745-R-00-003)

  This document is an addendum to the EPCRA section 313 Questions and Answers: Revised 1998 Version. It provides additional assistance to federal facilities in complying with EPCRA section 313. Federal facilities, which are subject to compliance under EPCRA through Executive Order 13423, frequently have operations that are different from the private sector facilities subject to EPCRA. The document contains questions and answers that address some of those differences.

- **EPCRA Section 313 Release and Other Waste Management Reporting Requirements**
  February 2001 (EPA 260/K-01-001)

  The brochure alerts businesses to their reporting obligations under EPCRA section 313 and assists in determining whether their facility is required to report. The brochure contains the EPA regional contacts, the list of EPCRA section 313 toxic chemicals and a description of the Standard Industrial Classification (SIC) codes subject to EPCRA section 313.

- **Toxic Chemical Release Reporting Using 2007 North American Industry Classification System (NAICS) Final Rule (73 FR 32466; June 9, 2008):** This final rule incorporates 2007 Office of Management and Budget (OMB) revisions and other corrections to the NAICS codes used for TRI Reporting.

- **Toxic Chemical Release Reporting Using North American Industry Classification System (NAICS) Final Rule (71 FR 32464; June 6, 2006):** With this rulemaking, Toxics Release Inventory (TRI) reporting will require North American Industry Classification System (NAICS) codes in place of Standard Industrial Classification

(SIC) codes. North American Industry Classification System (NAICS), United States, 2002, Executive Office of the President, Office of Management and Budget, NTIS Order Number: PB2002-101430

- **Persistent Bioaccumulative Toxic (PBT) Chemicals; Final Rule (64 FR 58666)**
  A reprint of the final rule that appeared in the *Federal Register* of October 29, 1999. This rule adds certain PBT chemicals and chemical categories for reporting year 2000 and beyond under EPCRA section 313, lowers their activity thresholds and modifies certain reporting exemptions and requirements for PBT chemicals and chemical categories. In a separate action, as part of the October 29, 1999 rulemaking, EPA added vanadium (except when contained in alloy) and vanadium compounds. These are not listed as PBT chemicals.

H.2 Supplier Notification Requirements

(EPA 560-4-91-006)

This pamphlet assists chemical suppliers who may be subject to the supplier notification requirements, gives examples of situations which require notification, describes the trade secret provision, and contains a sample notification.

- **Toxic Chemical Release Inventory Reporting Forms and Instructions Revised 2006 Version**
  February 2007 (EPA 260-C-06-901)

- **Toxics Release Inventory: Reporting Modifications Beginning with 1995 Reporting Year**
  February 1995 (EPA 745-R-95-009)

- **Trade Secrets Rule and Substantiation Form**
- **(53 FR 28772)**
  A reprint of the final rule that appeared in the *Federal Register* of July 29, 1988. This rule implements the trade secrets provision of the Emergency Planning and Community Right-to-Know Act (section 322). The current trade secret substantiation form can be accessed at http://www.epa.gov/tri/report/index.htm#forms

H.3 Chemical-Specific Guidance

EPA has developed a group of guidance documents specific to individual chemicals and chemical categories.

- **Emergency Planning and Community Right-to-Know Section 313: List of Toxic Chemicals within the Chlorophenols Category**
  June 1999 (EPA745-B-99-013)

- **Toxics Release Inventory List of Toxic Chemicals within the Glycol Ethers Category and Guidance for Reporting**
  December 2000 (EPA745-R-00-004)

- **Emergency Planning and Community Right-to-Know Act Section 313: Guidance for Reporting Hydrochloric Acid (acid aerosols including mists, vapors, gas, fog and other airborne forms of any particle size)**
  December 1999 (EPA 745-B-99-014)

- **Emergency Planning and Community Right-to-Know Act - Section 313: Guidance for Reporting Releases and Other Waste Management Activities of Toxic Chemicals: Lead and Lead Compounds**
  November 2001 (EPA-260-B-01-027)

- **Emergency Planning and Community Right-to-Know Act - Section 313: Guidance for Reporting Toxic Chemicals: Mercury and Mercury Compounds Category**
  August 2001 (EPA 260-B-01-004)

- **Toxics Release Inventory List of Toxic Chemicals within the Nicotine and Salt Category and Guidance for Reporting**
  June 1999 (EPA 745-R-99-010)

- **Toxics Release Inventory List of Toxic Chemicals within the Water Dissociable Nitrate Compounds Category and Guidance for Reporting**
  December 2000 (EPA 745-R-00-006)

- **Emergency Planning and Community Right-to-Know Act - Section 313: Guidance for Reporting Toxic Chemicals: Pesticides and Other Persistent Bioaccumulative Toxic (PBT) Chemicals**
  August 2001 (EPA 260-B-01-005)

- **Toxics Release Inventory List of Toxic Chemicals within the Polychlorinated Alkanes Category and Guidance for Reporting**
  June 1999 (EPA 745-B-99-023)

- **Emergency Planning and Community Right-to-Know Act - Section 313: Guidance for Reporting Toxic Chemicals: Polycyclic Aromatic Compounds Category**
  August 2001 (EPA 260-B-01-003)

- **Toxics Release Inventory List of Toxic Chemicals within the Strychnine and Salts Category and Guidance for Reporting**
  June 1999 (EPA 745-R-99-011)

- **Emergency Planning and Community Right-to-Know Act Section 313: Guidance for Reporting Sulfuric Acid** (acid aerosols including mists, vapors, gas, fog and other airborne forms of any particle size)
  March 1998 (EPA745-R-97-007)

- **Toxics Release Inventory List of Toxic Chemicals within Warfarin Category**
  June 1999 (EPA745-B-99-011)

- **Toxics Release Inventory List of Toxic Chemicals within Ethylenebisdithiocarbamic Acid, Salts and Esters Category and List of Mixtures that Contain the Individually listed Chemicals Maneb, Metiram, Nabam, and Zineb**
  September 2001 (EPA 260-B-01-026)

- **Emergency Planning and Community Right-to-Know Act - Section 313: Guidance for Reporting Aqueous Ammonia**
  December 2000 (EPA 745-R-00-005)

- **Emergency Planning and Community Right-to-Know Act - Section 313: Guidance for Reporting Toxic Chemicals within the Dioxin and Dioxin-like Compounds Category**
  December 2000 (EPA 745-B-00-021)

H.4 Industry-Specific Guidance

EPA has developed specific guidance documents for certain industries.

- **EPCRA Section 313: Guidance for Chemical Distribution Facilities**
  January 1999 (EPA 745-B-99-005)

- **EPCRA Section 313: Guidance for Petroleum Terminals and Bulk Storage Facilities**
  February 2000 (EPA 745-B-00-002)

- **EPCRA Section 313: Guidance for Coal Mining Facilities**
  February 2000 (EPA 745-B-00-003)

- **EPCRA Section 313: Guidance for Electricity Generating Facilities**
  February 2000 (EPA 745-B-00-004)

- **EPCRA Section 313 Reporting Guidance for Food Processors**

September 1998 (EPA 745-R-98-011)

- **EPCRA Section 313 Reporting Guidance for the Leather Tanning and Finishing Industry**
  April 2000 (EPA 745-B-00-012)

- **EPCRA Section 313: Guidance for Metal Mining Facilities**
  January1999 (EPA 745-B-99-001)

- **Emergency Planning and Community Right-to-Know Act Section 313 Reporting Guidance for the Presswood and Laminated Products Industry**
  August 2001 (EPA 260-B-01-013)

- **EPCRA Section 313 Reporting Guidance for the Printing, Publishing, and Packaging Industry**
  May 2000 (EPA 745-B-00-005)

- **EPCRA Section 313: Guidance for RCRA Subtitle C TSD Facilities and Solvent Recovery Facilities**
  January 1999 (EPA 745-B-99-004)

- **EPCRA Section 313 Reporting Guidance for Rubber and Plastics Manufacturing**
  May 2000 (EPA 745-B-00-017)

- **EPCRA Section 313 Reporting Guidance for Semiconductor Manufacturing**
  July 1999 (EPA 745-R-99-007)

- **EPCRA Section 313 Reporting Guidance for the Textile Processing Industry**
  May 2000 (EPA 745-B-00-008)

- **EPCRA Section 313 Reporting Guidance for Spray Application and Electrodeposition of Organic Coatings**
  December 1998 (EPA 745-R-98-014)

Appendix I. Questions and Answers Regarding Facility Identification Information

I.1 Categories

This document provides additional information about TRI reporting procedures based on some frequently asked questions. The questions and their answers are organized into three groups:

Section I.2   Identifying the parent company.
Section I.3   Reporting after a change in name or ownership.
Section I.4   Reporting for multiple sites and/or owners.

I.2 Identifying the Parent Company

A. Question
When a facility changes ownership after a Form R has been submitted, who is required to respond to a Notice of Noncompliance (NON) related to the Form R? Is the current or prior owner/operator required to respond to the NON?

A. Answer
The current owner/operator has the primary responsibility for responding to a NON.   However, all prior owners/operators back to January 1 of the reporting year may also be held responsible if the current owner/operator does not respond to the NON in an accurate, complete, and timely manner.
(Source: 1997 EPCRA Section 313 Questions and Answers Document, Question 31 (EPA 745-B-97-008)).

B. Question
Who is the parent company for a 50/50 joint venture?

B. Answer
The 50/50 joint venture is its own parent company.
(Source: 1997 EPCRA Section 313 Questions and Answers Document, Question #33 (EPA 745-B-97-008)).

C. Question
Mom and Pop Plastics is a wholly owned subsidiary of a major chemical company which is a wholly owned subsidiary of Big Oil Corporation, located in St. Paul, Minnesota. Which is the parent company?

C. Answer
Big Oil Corporation is the parent company.
(Source: 1997 EPCRA Section 313 Questions and Answers Document, Question #35 (EPA 745-B-97-008)).

I.3 Reporting After a Change in Name or Ownership

A. Question
The owner/operator of a covered facility is preparing Form Rs for a facility. The facility and its parent company both changed their names after the reporting year. What names should be reported by the owner/operator (for both the facility and the parent company) on the Form Rs covering the reporting year?

A. Answer
The facility should report the names used by the facility and parent company during that reporting year. When the owner/operator submits Form Rs for the next reporting year, these reports should reflect the names used by the facility and parent company during the new reporting year. Note that the TRI facility identification number will not change.
(Source: 1997 EPCRA Section 313 Questions and Answers Document, Question #457 (EPA 745-B-97-008)).

B. Question
If a covered facility does not have a Dun & Bradstreet (D&B) number but the parent corporation does, should this number be reported?

B. Answer
Report the D&B number for the facility. If a facility does not have a D&B number, enter "NA" in Part I, Section 4.7. The corporate D&B number should be entered in Part I, Section 5.2 relating to parent company information.
(Source: 1997 EPCRA Section 313 Questions and Answers Document, Question #464 (EPA 745-B-97-008)).

C. Question
In October 2009, Facility X changes ownership and is purchased by Company Y. For the 2009 reporting year, which facility is obligated to submit the Form R or Form A, and whose name and what TRI identification number should be on the form?

C. Answer
The owner or operator of the facility on the annual July 1 reporting deadline (i.e., Company Y) is primarily responsible for reporting the data for the entire previous

year's operations at that facility. Any other owner or operator of the facility before the reporting deadline may also be held liable. The form submitted for a given reporting year must reflect the names used by the facility and its parent company on December 31 of that reporting year, even if the facility changed its name or ownership at any time during the reporting year. In this scenario, because Facility X changed ownership before December 31 of the reporting year, Company Y's name should appear on the form. The TRI identification number is location-specific; thus, the identification number will stay the same even if the facility changes names, production processes, or NAICS codes.
(Source: Monthly Call Center Report Question EPA530-R-98---5j; October 1998).

I.4 Reporting for Multiple Sites and/or Owners

A. Question
If two plants are separate establishments under the same site management, must they have separate D&B numbers?

A. Answer
They may have separate D&B numbers, especially if they are distinctly separate business units. However, different divisions of a company located at the same facility usually do not have separate D&B numbers.
(Source: 1997 EPCRA Section 313 Questions and Answers Document, Question #465 (EPA 745-B-97-008)).

B. Question
An electricity generating facility (EGF) is comprised of multiple independent owners. Each individual owner runs his/her own separate operation, but each has a financial interest in the operation of the entire facility. What name should be entered as the parent company in Part I, Section 5.1 of the Form R? Should the facility report under one holding company name?

B. Answer
The EGF should enter in Part I, Section 5.1 of the Form R the name of the holding or parent company, consortium, joint venture, or other entity that owns, operates, or controls the facility.
(Source: Question #2, Addendum to the Guidance Documents for the Newly Added Industries (EPA 745-B-98-001)).

C. Question
A covered facility sells one of its establishments to a new owner. The operator of the newly sold establishment, however, does not change. The same operator operates the newly sold establishment and the rest of the facility. Although the facility makes its threshold determinations based on the activities at the entire facility (including the newly sold establishment), the facility chooses to report separately for the different establishments. What parent name should the newly sold establishment use, the parent name of the owner or the parent name of the operator (i.e., the same as the rest of the facility)?

C. Answer
All establishments of a covered facility must report the parent name of the facility. Therefore, in the instance described above, the newly sold establishment should use the parent name of the facility operator (i.e., the same parent name the rest of the facility is using).
(Source: Spring Training 1998).

D. Question
Company A purchases a facility from Company B between January 1, 2006 and June 30, 2006. For the 2005 reporting year, which company's name and identification number should appear on the Form R or Form A submission?

D. Answer
In the case that a facility is purchased between January 1 and June 30, the form submitted for the previous year must reflect the name used by the facility on December 31 of that reporting year. In this example, company B's name should appear on the form because it owned the facility for the duration of the reporting year. The TRI identification number is location-specific; thus, the identification number will stay the same even if the facility changes names, production processes, or NAICS codes.

With regard to reporting, the owner or operator of the facility on the annual July 1 reporting deadline (Company A) is primarily responsible for reporting the data for the previous year's operations at that facility. However, all prior owners and operators back to January 1 of the year covered in the report may also be held responsible if the current owner or operator does not submit a report.
(Source: Monthly Call Center Report Question EPA530-R-98---5j; October 1998)

E. Question

Two distinct NAICS code operations that are covered under EPCRA Section 313 (e.g., an electricity generating unit and a cement plant) are located on adjacent properties and are owned by the same parent company. The two operations are operated completely independently of one another (e.g., separate accounting procedures, employees, etc.). Are these two operations considered one facility under EPCRA Section 313?

E. Answer

Yes. Under EPCRA Section 313, a facility is defined as, "all buildings, equipment, structures, and other stationary items which are located on a single site or on contiguous or adjacent sites and which are owned or operated by the same person." Because these two operations are located on adjacent properties and are owned by the same person they are considered one facility for EPCRA Section 313 reporting purposes. Additional information can be found in the 2009 Toxic Release Inventory Reporting Forms and Instructions.

F. Question

A piece of contiguous property consists of three covered sites with various buildings, structures and equipment. The three sites are owned by two different companies – Company A and Company B. All three sites operate completely independently of each other and have separate personnel, finances, and environmental reporting systems. Site 1 and its buildings and structures are owned and operated by Company A and site 3 and its buildings and structures are owned and operated by Company B. The middle site, site 2 and its surrounding buildings and structures, are owned by Company A and operated by Company B. Are all three sites and their buildings and structures considered separate facilities under EPCRA Section 313? Who is responsible for reporting for each?

F. Answer

Under 40 CFR Section 372.3 a facility is defined as "all buildings, equipment, structures, and other stationary items which are located on a single site or on contiguous or adjacent sites and which are owned or operated by the same person." Because all buildings and structures located on sites 1 and 2 are located on contiguous property and are owned by the same person, they are considered one facility. Because all buildings and structures located on sites 2 and 3 are located on contiguous property and are operated by the same person, they are also considered one facility. Therefore, for purposes of determining thresholds, the toxic chemicals manufactured, processed, and otherwise used at site 2 must be counted toward both Facility A's and Facility B's threshold determinations. Because the operator is primarily responsible for reporting, estimating and reporting releases and other waste management calculations for sites 2 and 3 are the primary responsibility of Company B, and the release and other waste management reporting for site 1 is the primary responsibility of Company A. EPA allows the release and other waste management reporting to be done in this manner to avoid "double counting" releases and waste management activities at site 2. However, provided thresholds have been exceeded, if no reports are received from a covered facility, determinations can be found in the 2009 Toxic Release Inventory Reporting Forms and Instructions.